Spon's Estimating Costs Guide to
Minor Works, Refurbishment and Repairs

Spon's Contractors' Handbook Series

Prices, costs and estimates for all types of building works from £50 to £25,000 in value, specially written for contractors and small businesses.

Spon's Estimating Costs Guide to Plumbing and Heating
Spon's Estimating Costs Guide to Electrical Works
Spon's Estimating Costs Guide to Minor Works, Refurbishment and Repairs

Spon's Price Book Series

For larger building works, covering all types and sizes of construction contracts. The 2002 editions, with free CD-ROM, are out now.

Spon's Architects' and Builders' Price Book
Spon's Mechanical and Electrical Services Price Book
Spon's Civil Engineering and Highway Works Price Book
Spon's Landscape and External Works Price Book

For further details on these, and all other Spon publications, please visit www.sponpress.com or www.pricebooks.co.uk

Spon's Estimating Costs Guide to Minor Works, Refurbishment and Repairs

Bryan Spain

Spon's Contractors' Handbooks

London and New York

First published 2002 by Spon Press
11 New Fetter Lane, London EC4P 4EE

Simultaneously published in the USA and Canada
by Spon Press
29 West 35th Street, New York, NY 10001

Reprinted 2003 (twice)

Spon Press is an imprint of the Taylor & Francis Group

Publisher's note

This book has been prepared from camera-ready copy provided by the author.

Printed and bound in Great Britain by TJ International Ltd, Padstow, Cornwall.

British Library Cataloguing in Publication Data
A catalogue record for this book is available from the British Library

Library of Congress Cataloging in Publication Data
Spain, Bryan J. D.
        Spon's estimating costs guide to minor works, refurbishment and repair/ Bryan Spain
        p. cm. -- (Spon's contractors' handbooks)
        Include index.
        1. Buildings--Repairs and reconstruction--Estimates. 2
        Buildings--Maintenance--Estimates. 3. Building--Costs. I. Title. II. Series

        TH435 .S725372001
        692'.5--dc21   2001049019

ISBN 0-415-26645-9

Contents

Preface   xiii

Introduction   xv

Standard Method of Measurement/trades link   xix

Part One: Unit rates

Demolition, excavation and filling

    Demolition   3
    Excavation   5
    Earthwork support   7
    Disposal by hand   7
    Disposal by machine   8
    Filling by hand   9
    Filling by machine   9
    Surface treatments   10

Concrete work

    Ready-mixed concrete   11
    Site-mixed concrete   12
    Formwork   13
    Reinforcement   17
    Joints   18
    Concrete finishes   18
    Precast concrete   18

Brickwork and blockwork

    Brickwork   20
    Blockwork   30
    Damp-proof courses   34
    Sundries   36

Masonry

Walling	37
Sundries	39

Carpentry and joinery

Sawn softwood, untreated	41
Sawn softwood, impregnated	42
Gutters and fascias	44
Wrought softwood supports	45
Metal fixings	46
Flooring	48
Linings	49
Windows	50
Doors	52
Door frames and linings	53
Stairs	55
Kitchen fittings	56
Insulation	57
Ironmongery	57

Metalwork

Balustrades	60
Lintels	60
Sundries	61

Roofing

Tiling	62
Fibre-cement slating	65
Natural slating	66
Reconstructed stone slating	68
Lead sheet coverings	69
Copper sheet coverings	72
Built-up roofing	73

Asphalt work

Damp proofing and tanking	75
Flooring	76
Roofing	77

Floor, wall and ceiling finishes

Screeds 79
In situ wall coverings 83
Lathing 85
Plasterboard 86
Floor tiling 86
Wall tiling 88

Plumbing and heating 90

Sanitary fittings 90
Holes and chases 92
Rainwater pipes and gutters 94
Waste pipes 96
Soil pipes 97
Overflows 99
Traps 99
Copper pipework, capillary joints 100
Stop valves 104
Valves 104
Copper pipework, compression joints 104
Stop valves 106
Cold water storage tanks 106
Hot water tanks 107
Hot water copper cylinders 107
Insulation 109
Boilers gas-fired 110
Oil storage tanks 112
Radiators 114

Glazing

Clear float glass 117
White patterned glass 119
Georgian wired cast glass 121
Georgian wired polished glass 122
Clear laminated safety glass 123

Painting and wallpapering

On internal surfaces before fixing 125
On internal walls and ceilings 125
On internal woodwork 127
On internal metalwork 132

Painting and wallpapering (cont'd)

On external woodwork 136
On external metalwork 141
Wallpapering 146

External works

Drainage 148
Fencing 157
Kerbs and edgings 159
Sub-bases 160
Concrete beds 161
Pavings 162

Alterations and repairs

Shoring 164
Forming openings 165
Filling openings 167
Underpinning 168
Temporary screens 173

Spot items

Brickwork, blockwork and masonry 173
Roofing 177
Carpentry and joinery 180
Finishings 182
Plumbing 183
Glazing 185
Painting 186
Wallpapering 186

Part Two: Approximate estimating

Excavation and filling 191
Concrete work 192
Brickwork and blockwork 193
Masonry 194
Carpentry and joinery 195
Finishings 198
Plumbing and heating 198

Approximate estimating (cont'd)

Painting	200
Wallpapering	201
Drainage	202

Part Three: Plant and tool hire

Tool and equipment hire

Concrete mixers	207
Vibrating pokers	207
Power floats	207
Vibrating screeds	208
Floor preparation units	208
Floor saws	208
Disc cutters	208
Block and slab splitters	209
Saw benches	209

Excavating and site plant

Excavators	209
Air compressors	209
Loaders and carriers	209
Dumpers	209

Access and site equipment

Ladders	210
Props	210
Chimney scaffold	210
Rubbish chutes	211
Trestles	211
Staging boards	211
Alloy towers	211

Material handling

Lifting and moving	212

Compaction

    Plate compactors   212
    Rammers   213
    Vibrating rollers   213

Breaking and demolition

    Breakers   213
    Hydraulic breakers   213

Power tools

    Drills   213
    Impact wrenches   214
    Grinders   214
    Saws   214
    Woodworking   214
    Sanders   215

Welding and generators

    Generating   215

Pumping equipment

    Pumps   215

Painting and decorating

    Surface preparation   216
    Damp proofing   216

Part Four: General construction data

General construction data   219

Part Five: Business matters

Starting a business   239

Running a business   253

Taxation   259

Index   269

Preface

The construction industry has never been under such scrutiny as it is today. Stringent financial constraints apply to both the public and private sectors whilst the apparently insatiable appetite of the general public for TV shows on 'make-overs' has produced a new breed of 'experts'.

Yet the small building contractor continues to provide an often unappreciated service to the community in spite of a few unscrupulous characters whose activities damage the reputation of hardworking, honest individuals and firms.

Carrying out building work is often the simplest part of a small contractor's workload. Finding time to fill in forms, prepare estimates, chase slow payers, order materials and carry out other tasks that do not produce a direct financial return is becoming more difficult and can often make the difference between a firm making a profit or a loss.

This book is intended to help contractors to prepare their estimates by providing thousands of rates for small building work. If used sensibly, the book can save valuable time for contractors in costing their own work, and can also play a part in making sure that value for money is received from sub-contractors.

I have received a great deal of support in the research necessary for this type of book and I am grateful to those individuals and firms who have provided the cost data and other information. In particular, I am indebted to Mark Loughrey of Youds, Ellison & Co., Chartered Accountants of Hoylake (tel: 0151-632 3298 or www.yesl.uk.com), who are specialists in advising small businesses. Their research for the information in the business section is based on tax legislation in force in December 2000.

Although every care has been taken in the preparation of the book, neither the publishers nor I can accept any responsibility for the use of the information provided by any firm or individual. Finally, I would welcome any constructive criticism of the book's contents and suggestions that could be incorporated into future editions.

Bryan Spain
August 2001

Introduction

The contents of this book cover unit rates, general advice on business matters and other information useful to those involved in the commissioning and construction of building work. The unit rates section presents analytical rates for work up to about £50,000. Rates for alteration and repair work are also included.

The business section covers advice on starting and running a business together with information on taxation and VAT matters.

Materials

As the users of this book will mainly be small contractors, it has been assumed that they will probably not receive the benefits of the discounts available to contractors who are able to buy materials in large quantities. An average of 10% to 15% discount has been allowed on normal trade prices.

Labour

The net labour rates for craftsmen and general operatives are based upon the wage awards promulgated by the Building and Allied Trades Joint Industrial Council (BATJIC) effective from 11 June 2001.

			Craftsman £
Flat time 48 weeks @ £261.3			12,542.40
Bonus 48 weeks @ £25.00			1,200.00
		£	13,742.40
NIC Employers' contribution	12.2%		1,676.57
B and CE Benefits scheme	10%		1,374.24
CITB levy	0.25%		34.36
Carried forward			16,827.57

Brought forward		16,827.57
Employers' liability	2%	274.85
Non-productive overtime 39 hours @ £6.70		261.30
Public holidays 67.5 hours @ £6.70		452.25
		17,815.97
divided by 1,762 productive hours		10.11

		General Operative £
Flat time 48 weeks @ £211.38		10,146.24
Bonus 48 weeks @ £15.00		720.00
		10,866.24
NIC Employers' contribution	12.2%	1,325.68
B and CE Benefits scheme	10%	1,086.62
CITB levy	0.25%	27.17
Employers' liability	2%	217.32
Non-productive overtime 39 hours @ £5.42		211.38
Public holidays 67.5 hours @ £5.42		365.85
		14,100.26
divided by 1,762 productive hours		£8.00

The labour rate for plumbing work has been taken as £10.94 per hour based upon the Joint Industry Board for Plumbing Mechanical Services (JIBPMES) payable to Advanced Plumbers from 24 August 2000, including a £25 weekly bonus and employer's costs.

Overheads and profit

This has been set at 15% for all grades of work and is deemed to cover head office and site overheads including:

- heating
- lighting
- rent
- rates
- telephones
- secretarial services
- insurances
- finance charges
- transport
- small tools
- ladders
- scaffolding etc.

Standard Method of Measurement/trades link

The contents of this book are presented under trade headings and the following table provides a link to the Standard Method of Measurement (SMM).

Demolition, excavation and filling

    C10 Demolishing structures
    D20 Excavation and filling

Concrete work

    E10 In situ concrete
    E20 Formwork for in situ concrete
    E30 Reinforcement for in situ concrete
    E40 Designed joints for in situ concrete
    E41 Worked finished to situ concrete
    E50 Precast concrete units

Brickwork and blockwork

    F10 Brick/block walling
    F30 Accessories for brick/block walling

Masonry

    F20 Natural stone rubble walling
    F30 Accessories for stone walling

Carpentry and joinery

    G20 Carpentry/timber framing/first fixing
    G32 Edge supported woodwool decking
    K20 Timber board flooring
    L10 Timber windows
    L20 Timber doors
    L30 Timber stairs
    N11 Domestic fittings

Carpentry and joinery (cont'd)

> Y50 Thermal insulation
> P21 Ironmongery

Metalwork

> L31 Balustrades

Roofing

> H60 Clay/concrete roof tiling
> H61 Fibre-cement slating
> H62 Natural slating
> H63 Reconstructed stone slating
> H71 Lead sheet coverings
> H73 Copper sheet coverings
> J41 Built-up roofing

Asphalt work

> J20 Mastic asphalt tanking
> M11 Mastic asphalt flooring
> J21 Mastic asphalt roofing

Floor, wall and ceiling finishes

> M10 Sand cement/granolithic screeds
> M20 Plastered coatings
> M30 Metal mesh lathing
> K10 Plasterboard dry lining
> M40 Quarry/ceramic tiling
> M50 Plastic/lino tiling

Plumbing and heating

> N13 Sanitary fittings
> R10 Rainwater pipes and gutters
> R11 Foul drainage above ground
> S10 Piped supply systems
> Y21 Water tank/cisterns
> Y23 Storage cylinders
> S41 Fuel oil storage
> T31 Low temperature hot water heating

Glazing

L40 General glazing

Painting and wallpapering

M60 Painting/clear finishing
M52 Decorative papers
Internal metalwork
Wallpapering

External works

R12 Drainage below ground
Q40 Fencing
Q10 Concrete kerbs and edgings
Q20 Hardcore/granular sub-bases
Q25 Slab/brick pavings

Alterations and repairs

C30 Shoring
D50 Underpinning
C10 Demolishing structures
C20 Spot items

Part One

UNIT RATES

Demolition, excavation and filling

Concrete work

Brickwork and blockwork

Masonry

Carpentry and joinery

Metalwork

Roofing

Asphalt work

Floor, wall and ceiling finishings

Plumbing and heating

Glazing

Painting and wallpapering

External works

Alterations and repairs

Spot items

	Unit	Hours	Hours £	Materials £	O & P £	Total £

DEMOLITION, EXCAVATION AND FILLING

Demolition

Demolish ground level existing brick buildings

	Unit	Hours	Hours £	Materials £	O & P £	Total £
single storey, detached	m3	1.60	12.80	0.00	1.92	14.72
single storey, attached	m3	1.70	13.60	0.00	2.04	15.64
two storey, attached	m3	1.80	14.40	0.00	2.16	16.56

Break up debris and arisings, load into skips or lorries

unreinforced concrete slabs

	Unit	Hours	Hours £	Materials £	O & P £	Total £
100mm thick	m2	0.80	6.40	0.00	0.96	7.36
150mm thick	m2	1.00	8.00	0.00	1.20	9.20
200mm thick	m2	1.40	11.20	0.00	1.68	12.88

reinforced concrete slabs

	Unit	Hours	Hours £	Materials £	O & P £	Total £
100mm thick	m2	1.40	11.20	0.00	1.68	12.88
150mm thick	m2	2.00	16.00	0.00	2.40	18.40
200mm thick	m2	2.60	20.80	0.00	3.12	23.92

reinforced suspended concrete slabs

	Unit	Hours	Hours £	Materials £	O & P £	Total £
100mm thick	m2	2.80	22.40	0.00	3.36	25.76
150mm thick	m2	3.20	25.60	0.00	3.84	29.44
200mm thick	m2	3.40	27.20	0.00	4.08	31.28

reinforced concrete beams

	Unit	Hours	Hours £	Materials £	O & P £	Total £
200 x 300mm	m	1.25	10.00	0.00	1.50	11.50
250 x 350mm	m	1.50	12.00	0.00	1.80	13.80
300 x 400mm	m	1.75	14.00	0.00	2.10	16.10

reinforced concrete columns

	Unit	Hours	Hours £	Materials £	O & P £	Total £
200 x 300mm	m	1.30	10.40	0.00	1.56	11.96
250 x 350mm	m	1.60	12.80	0.00	1.92	14.72
300 x 400mm	m	1.90	15.20	0.00	2.28	17.48

	Unit	Hours	Hours £	Materials £	O & P £	Total £
Demolish brick walls in cement mortar						
half brick thick	m2	0.80	6.40	0.00	0.96	7.36
one brick thick	m2	1.40	11.20	0.00	1.68	12.88
one and a half brick thick	m2	1.90	15.20	0.00	2.28	17.48
two brick thick	m2	2.70	21.60	0.00	3.24	24.84
Demolish brick walls in cement lime mortar						
half brick thick	m2	0.70	5.60	0.00	0.84	6.44
one brick thick	m2	1.30	10.40	0.00	1.56	11.96
one and a half brick thick	m2	1.80	14.40	0.00	2.16	16.56
two brick thick	m2	2.50	20.00	0.00	3.00	23.00
Carefully take down brick walls in cement lime mortar, clean bricks and lay aside for re-use						
half brick thick	m2	1.00	8.00	0.00	1.20	9.20
one brick thick	m2	1.70	13.60	0.00	2.04	15.64
one and a half brick thick	m2	2.20	17.60	0.00	2.64	20.24
two brick thick	m2	3.20	25.60	0.00	3.84	29.44
Take out fireplace and surround, load into skips or lorries						
1500 x1200mm high	nr	1.50	12.00	0.00	1.80	13.80
1500 x1500mm high	nr	1.75	14.00	0.00	2.10	16.10
1800 x1500mm high	nr	2.00	16.00	0.00	2.40	18.40
Demolish blockwork partitions						
75mm thick	m2	0.80	6.40	0.00	0.96	7.36
100mm thick	m2	0.90	7.20	0.00	1.08	8.28
115mm thick	m2	1.00	8.00	0.00	1.20	9.20
140mm thick	m2	1.10	8.80	0.00	1.32	10.12
190mm thick	m2	1.20	9.60	0.00	1.44	11.04
215mm thick	m2	1.40	11.20	0.00	1.68	12.88

	Unit	Hours	Hours £	Plant £	O & P £	Total £

Excavation

Excavate by hand

topsoil
	Unit	Hours	Hours £	Plant £	O & P £	Total £
150mm thick	m2	0.35	2.80	0.00	0.42	3.22
200mm thick	m2	0.45	3.60	0.00	0.54	4.14
250mm thick	m2	0.60	4.80	0.00	0.72	5.52

to reduce levels, depth not exceeding
	Unit	Hours	Hours £	Plant £	O & P £	Total £
0.25m	m3	2.20	17.60	0.00	2.64	20.24
1.00m	m3	2.40	19.20	0.00	2.88	22.08
1.50m	m3	2.80	22.40	0.00	3.36	25.76
2.00m	m3	3.20	25.60	0.00	3.84	29.44

to trenches 450mm wide, depth not exceeding
	Unit	Hours	Hours £	Plant £	O & P £	Total £
0.25m	m3	2.50	20.00	0.00	3.00	23.00
1.00m	m3	2.70	21.60	0.00	3.24	24.84
1.50m	m3	3.00	24.00	0.00	3.60	27.60
2.00m	m3	3.40	27.20	0.00	4.08	31.28

to pits, depth not exceeding
	Unit	Hours	Hours £	Plant £	O & P £	Total £
0.25m	m3	2.50	20.00	0.00	3.00	23.00
1.00m	m3	2.70	21.60	0.00	3.24	24.84
1.50m	m3	3.10	24.80	0.00	3.72	28.52
2.00m	m3	3.50	28.00	0.00	4.20	32.20

Extra for excavating through
	Unit	Hours	Hours £	Plant £	O & P £	Total £
rock	m3	10.00	80.00	0.00	12.00	92.00
concrete	m3	8.00	64.00	0.00	9.60	73.60
reinforced concrete	m3	9.00	72.00	0.00	10.80	82.80
brickwork	m3	6.00	48.00	0.00	7.20	55.20

	Unit	Hours	Hours £	Plant £	O & P £	Total £
Excavate by machine						
topsoil						
150mm thick	m2	0.04	0.32	1.20	0.23	1.75
200mm thick	m2	0.05	0.40	1.30	0.26	1.96
250mm thick	m2	0.06	0.48	1.40	0.28	2.16
to reduce levels, depth not exceeding						
0.25m	m3	0.10	0.80	1.40	0.33	2.53
1.00m	m3	0.08	0.64	1.20	0.28	2.12
1.50m	m3	0.10	0.80	1.40	0.33	2.53
2.00m	m3	0.12	0.96	1.60	0.38	2.94
to trenches 450mm wide, depth not exceeding						
0.25m	m3	0.35	2.80	4.00	1.02	7.82
1.00m	m3	0.30	2.40	3.80	0.93	7.13
1.50m	m3	0.35	2.80	4.00	1.02	7.82
2.00m	m3	0.40	3.20	4.50	1.16	8.86
to pits, depth not exceeding						
0.25m	m3	0.40	3.20	4.50	1.16	8.86
1.00m	m3	0.35	2.80	4.20	1.05	8.05
1.50m	m3	0.40	3.20	4.50	1.16	8.86
2.00m	m3	0.45	3.60	4.80	1.26	9.66
Extra for excavating through						
rock	m3	5.00	40.00	16.00	8.40	64.40
concrete	m3	3.75	30.00	12.50	6.38	48.88
reinforced concrete	m3	4.00	32.00	14.30	6.95	53.25
brickwork	m3	3.80	30.40	11.50	6.29	48.19

	Unit	Hours	Hours £	Materials £	O & P £	Total £

Earthwork support

Earthwork support not exceeding
2m between opposite faces, depth
not exceeding 1m

	Unit	Hours	Hours £	Materials £	O & P £	Total £
firm ground	m2	0.15	1.20	1.05	0.34	2.59
loose ground	m2	0.85	6.80	2.25	1.36	10.41
sand	m2	1.10	8.80	2.80	1.74	13.34

Earthwork support not exceeding
2m between opposite faces, depth
not exceeding 2m

	Unit	Hours	Hours £	Materials £	O & P £	Total £
firm ground	m2	0.18	1.44	1.05	0.37	2.86
loose ground	m2	0.90	7.20	2.25	1.42	10.87
sand	m2	1.20	9.60	2.80	1.86	14.26

Earthwork support not exceeding
2m between opposite faces, depth
not exceeding 4m

	Unit	Hours	Hours £	Materials £	O & P £	Total £
firm ground	m2	0.22	1.76	1.05	0.42	3.23
loose ground	m2	0.95	7.60	2.25	1.48	11.33
sand	m2	1.30	10.40	2.80	1.98	15.18

Disposal by hand

Load surplus excavated material
into barrows, wheel and deposit
in temporary spoil heaps, average
distance

	Unit	Hours	Hours £	Materials £	O & P £	Total £
15m	m3	1.25	10.00	0.00	1.50	11.50
25m	m3	1.45	11.60	0.00	1.74	13.34
50m	m3	1.70	13.60	0.00	2.04	15.64

	Unit	Hours	Hours £	Plant £	O & P £	Total £
Load surplus excavated material into barrows, wheel and spread over site, average distance						
15m	m3	1.65	13.20	0.00	1.98	15.18
25m	m3	1.85	14.80	0.00	2.22	17.02
50m	m3	2.00	16.00	0.00	2.40	18.40
Load surplus excavated material into barrows, wheel and deposit in skips or lorries, average distance						
15m	m3	1.20	9.60	0.00	1.44	11.04
25m	m3	1.40	11.20	0.00	1.68	12.88
50m	m3	1.65	13.20	0.00	1.98	15.18

Disposal by machine

	Unit	Hours	Hours £	Plant £	O & P £	Total £
Load and deposit in temporary spoil heaps, average distance						
15m	m3	0.10	0.80	0.90	0.26	1.96
25m	m3	0.15	1.20	1.10	0.35	2.65
50m	m3	0.25	2.00	1.60	0.54	4.14
Load and spread over site, average distance						
15m	m3	0.15	1.20	1.10	0.35	2.65
25m	m3	0.18	1.44	1.25	0.40	3.09
50m	m3	0.24	1.92	1.40	0.50	3.82
Load and deposit in skips or lorries, average distance						
15m	m3	0.08	0.64	0.90	0.23	1.77
25m	m3	0.12	0.96	1.10	0.31	2.37
50m	m3	0.22	1.76	1.60	0.50	3.86

	Unit	Hours	Hours £	Materials £	O & P £	Total £
Filling by hand						
Surplus excavated material deposited and compacting in layers						
over 250mm thick	m3	1.20	9.60	2.66	1.84	14.10
100mm thick	m2	0.20	1.60	0.40	0.30	2.30
150mm thick	m2	0.35	2.80	0.50	0.50	3.80
200mm thick	m2	0.50	4.00	0.60	0.69	5.29
Imported sand deposited and compacting in layers						
over 250mm thick	m3	1.20	9.60	19.45	4.36	33.41
100mm thick	m2	0.20	1.60	2.30	0.59	4.49
150mm thick	m2	0.35	2.80	3.24	0.91	6.95
200mm thick	m2	0.50	4.00	4.55	1.28	9.83
Imported hardcore deposited and compacting in layers						
over 250mm thick	m3	1.20	9.60	16.44	3.91	29.95
100mm thick	m2	0.20	1.60	2.10	0.56	4.26
150mm thick	m2	0.35	2.80	2.86	0.85	6.51
200mm thick	m2	0.50	4.00	3.65	1.15	8.80
Filling by machine						
Surplus excavated material deposited and compacting in layers						
over 250mm thick	m3	0.40	3.20	2.66	0.88	6.74
100mm thick	m2	0.06	0.48	0.40	0.13	1.01
150mm thick	m2	0.08	0.64	0.50	0.17	1.31
200mm thick	m2	0.10	0.80	0.60	0.21	1.61

	Unit	Hours	Hours £	Plant £	O & P £	Total £
Imported sand deposited and compacting in layers						
over 250mm thick	m3	0.40	3.20	19.45	3.40	26.05
100mm thick	m2	0.06	0.48	2.30	0.42	3.20
150mm thick	m2	0.08	0.64	3.24	0.58	4.46
200mm thick	m2	0.10	0.80	4.55	0.80	6.15
Imported hardcore deposited and compacting in layers						
over 250mm thick	m3	0.40	3.20	16.44	2.95	22.59
100mm thick	m2	0.06	0.48	2.10	0.39	2.97
150mm thick	m2	0.08	0.64	2.86	0.53	4.03
200mm thick	m2	0.10	0.80	3.65	0.67	5.12

Surface treatments

	Unit	Hours	Hours £	Plant £	O & P £	Total £
Level and compact bottom of excavation with vibrating roller	m2	0.10	0.80	3.65	0.67	5.12
Blind filling surfaces with sand 50mm thick	m2	0.12	0.96	1.20	0.32	2.48

	Unit	Hours	Hours £	Materials £	O & P £	Total £

CONCRETE WORK

Ready-mixed concrete 1:3:6
40mm aggregate

Foundations in trenches, thickness

150 to 300mm	m3	1.85	14.80	67.94	12.41	95.15
300 to 450mm	m3	1.35	10.80	67.94	11.81	90.55
over 450mm	m3	1.15	9.20	67.94	11.57	88.71

Isolated column bases

150 to 300mm	m3	2.00	16.00	67.94	12.59	96.53
300 to 450mm	m3	1.50	12.00	67.94	11.99	91.93
over 450mm	m3	1.30	10.40	67.94	11.75	90.09

Filling to cavity walls	m3	3.95	31.60	67.94	14.93	114.47

Ready-mixed concrete 1:2:4
20mm aggregate

Beds

150 to 300mm	m3	2.40	19.20	72.56	13.76	105.52
300 to 450mm	m3	2.10	16.80	72.56	13.40	102.76
over 450mm	m3	1.80	14.40	72.56	13.04	100.00

Slabs

150 to 300mm	m3	3.20	25.60	72.56	14.72	112.88
300 to 450mm	m3	2.85	22.80	72.56	14.30	109.66

Walls

150 to 300mm	m3	4.25	34.00	72.56	15.98	122.54
300 to 450mm	m3	3.95	31.60	72.56	15.62	119.78

	Unit	Hours	Hours £	Materials £	O & P £	Total £
Casings						
isolated beams	m3	5.00	40.00	72.56	16.88	129.44
isolated deep beams	m3	4.50	36.00	72.56	16.28	124.84
attached deep beams	m3	5.00	40.00	72.56	16.88	129.44
columns	m3	5.00	40.00	72.56	16.88	129.44
staircases	m3	6.00	48.00	72.56	18.08	138.64
Upstands and kerbs	m3	4.40	35.20	72.56	16.16	123.92
Site-mixed concrete 1:3:6 40mm aggregate						
Foundations in trenches, thickness						
150 to 300mm	m3	2.85	22.80	62.15	12.74	97.69
300 to 450mm	m3	2.35	18.80	62.15	12.14	93.09
over 450mm	m3	2.15	17.20	62.15	11.90	91.25
Isolated column bases						
150 to 300mm	m3	3.00	24.00	62.15	12.92	99.07
300 to 450mm	m3	2.50	20.00	62.15	12.32	94.47
over 450mm	m3	2.30	18.40	62.15	12.08	92.63
Filling to cavity walls	m3	4.95	39.60	62.15	15.26	117.01
Site-mixed concrete 1:2:4 20mm aggregate						
Beds						
150 to 300mm	m3	3.40	27.20	69.47	14.50	111.17
300 to 450mm	m3	3.10	24.80	69.47	14.14	108.41
over 450mm	m3	1.90	15.20	69.47	12.70	97.37
Slabs						
150 to 300mm	m3	4.20	33.60	69.47	15.46	118.53
300 to 450mm	m3	3.85	30.80	69.47	15.04	115.31

	Unit	Hours	Hours £	Materials £	O & P £	Total £
Walls						
150 to 300mm	m3	5.25	42.00	69.47	16.72	128.19
300 to 450mm	m3	4.95	39.60	69.47	16.36	125.43
Casings						
isolated beams	m3	6.00	48.00	69.47	17.62	135.09
isolated deep beams	m3	5.50	44.00	69.47	17.02	130.49
attached deep beams	m3	6.00	48.00	69.47	17.62	135.09
columns	m3	6.00	48.00	69.47	17.62	135.09
staircases	m3	7.00	56.00	69.47	18.82	144.29
Upstands and kerbs	m3	5.50	44.00	69.47	17.02	130.49

Formwork

Plain vertical to sides of foundations

	Unit	Hours	Hours £	Materials £	O & P £	Total £
over 1m	m2	2.30	18.40	7.98	3.96	30.34
not exceeding 250mm	m	0.75	6.00	2.80	1.32	10.12
250 to 500mm	m	1.25	10.00	4.85	2.23	17.08
500mm to 1m	m	1.70	13.60	7.98	3.24	24.82

Plain vertical to sides of
foundations, left in

	Unit	Hours	Hours £	Materials £	O & P £	Total £
over 1m	m2	2.15	17.20	19.40	5.49	42.09
not exceeding 250mm	m	0.65	5.20	6.90	1.82	13.92
250 to 500mm	m	1.15	9.20	12.76	3.29	25.25
500mm to 1m	m	1.60	12.80	19.40	4.83	37.03

Plain vertical to sides of ground
beams and beds

	Unit	Hours	Hours £	Materials £	O & P £	Total £
over 1m	m2	2.40	19.20	7.98	4.08	31.26
not exceeding 250mm	m	0.82	6.56	2.80	1.40	10.76
250 to 500mm	m	1.34	10.72	4.85	2.34	17.91
500mm to 1m	m	1.78	14.24	7.98	3.33	25.55

	Unit	**Hours**	**Hours £**	**Materials £**	**O & P £**	**Total £**
Plain vertical to sides of ground beams and beds, left in						
over 1m	m2	2.25	18.00	19.40	5.61	43.01
not exceeding 250mm	m	0.75	6.00	6.90	1.94	14.84
250 to 500mm	m	1.25	10.00	12.76	3.41	26.17
500mm to 1m	m	1.70	13.60	19.40	4.95	37.95
Edges of suspended slabs						
not exceeding 250mm	m	0.95	7.60	2.80	1.56	11.96
250 to 500mm	m	1.40	11.20	4.85	2.41	18.46
Sides of upstands						
over 1m	m2	2.60	20.80	7.98	4.32	33.10
not exceeding 250mm	m	1.00	8.00	2.80	1.62	12.42
250 to 500mm	m	1.50	12.00	4.85	2.53	19.38
500mm to 1m	m	1.90	15.20	7.98	3.48	26.66
Steps in top surfaces						
not exceeding 250mm	m	0.75	6.00	2.80	1.32	10.12
250 to 500mm	m	1.25	10.00	4.85	2.23	17.08
500mm to 1m	m	1.70	13.60	7.98	3.24	24.82
Steps in soffits						
not exceeding 250mm	m	1.00	8.00	2.80	1.62	12.42
250 to 500mm	m	1.50	12.00	4.85	2.53	19.38
500mm to 1m	m	1.90	15.20	7.98	3.48	26.66
Machine bases and plinths						
over 1m	m2	2.30	18.40	7.98	3.96	30.34
not exceeding 250mm	m	0.75	6.00	2.80	1.32	10.12
250 to 500mm	m	1.25	10.00	4.85	2.23	17.08
500mm to 1m	m	1.70	13.60	7.98	3.24	24.82

	Unit	Hours	Hours £	Materials £	O & P £	Total £
Horizontal to soffits of slabs, thickness						
not exceeding 200mm						
height to soffit, not exceeding						
1.5m	m2	2.20	17.60	7.98	3.84	29.42
height to soffit, 1.5 to 3m	m2	2.10	16.80	7.98	3.72	28.50
height to soffit, over 3m	m2	2.00	16.00	7.98	3.60	27.58
200 to 300mm						
height to soffit, not exceeding						
1.5m	m2	2.40	19.20	7.98	4.08	31.26
height to soffit, 1.5 to 3m	m2	2.30	18.40	7.98	3.96	30.34
height to soffit, over 3m	m2	2.20	17.60	7.98	3.84	29.42
300 to 400mm						
height to soffit, not exceeding						
1.5m	m2	2.60	20.80	7.98	4.32	33.10
height to soffit, 1.5 to 3m	m2	2.50	20.00	7.98	4.20	32.18
height to soffit, over 3m	m2	2.40	19.20	7.98	4.08	31.26
Horizontal to soffits of slabs, left in, thickness						
not exceeding 200mm						
height to soffit, not exceeding						
1.5m	m2	2.25	18.00	19.40	5.61	43.01
height to soffit, 1.5 to 3m	m2	2.20	17.60	19.40	5.55	42.55
height to soffit, over 3m	m2	2.10	16.80	19.40	5.43	41.63
200 to 300mm						
height to soffit, not exceeding						
1.5m	m2	2.35	18.80	19.40	5.73	43.93
height to soffit, 1.5 to 3m	m2	2.30	18.40	19.40	5.67	43.47
height to soffit, over 3m	m2	2.20	17.60	19.40	5.55	42.55

	Unit	Hours	Hours £	Materials £	O & P £	Total £
Horizontal to soffits of slabs, left in, thickness (cont'd)						
300 to 400mm height to soffit, not exceeding						
1.5m	m2	2.45	19.60	19.40	5.85	44.85
height to soffit, 1.5 to 3m	m2	2.40	19.20	19.40	5.79	44.39
height to soffit, over 3m	m2	2.30	18.40	19.40	5.67	43.47
Walls						
vertical	m2	2.40	19.20	7.98	4.08	31.26
vertical interrupted	m2	2.50	20.00	7.98	4.20	32.18
Beams						
attached to slabs, height to soffit						
1.5 to 3m	m2	3.20	25.60	7.98	5.04	38.62
3 to 4.5m	m2	3.30	26.40	7.98	5.16	39.54
over 4.5m	m2	3.50	28.00	7.98	5.40	41.38
Columns						
attached to walls, height to soffit						
1.5 to 3m	m2	3.20	25.60	7.98	5.04	38.62
3 to 4.5m	m2	3.30	26.40	7.98	5.16	39.54
over 4.5m	m2	3.50	28.00	7.98	5.40	41.38
isolated, height to soffit						
1.5 to 3m	m2	3.40	27.20	7.98	5.28	40.46
3 to 4.5m	m2	3.50	28.00	7.98	5.40	41.38
over 4.5m	m2	3.60	28.80	7.98	5.52	42.30
Mortice in concrete for rag bolt, grout in cement mortar (1:3), depth						
50mm	nr	0.30	2.40	0.80	0.48	3.68
100mm	nr	0.35	2.80	1.75	0.68	5.23
150mm	nr	0.40	3.20	2.88	0.91	6.99

	Unit	Hours	Hours £	Materials £	O & P £	Total £

Reinforcement

Plain round steel reinforcement
bars, straight or bent

6mm diameter	m	0.02	0.16	0.14	0.05	0.35
8mm diameter	m	0.03	0.24	0.20	0.07	0.51
10mm diameter	m	0.04	0.32	0.28	0.09	0.69
12mm diameter	m	0.05	0.40	0.36	0.11	0.87
16mm diameter	m	0.06	0.48	0.56	0.16	1.20
20mm diameter	m	0.07	0.56	0.90	0.22	1.68
25mm diameter	m	0.08	0.64	1.44	0.31	2.39

High yield deformed steel
reinforcement bars, straight or bent

8mm diameter	m	0.03	0.24	0.18	0.06	0.48
10mm diameter	m	0.04	0.32	0.26	0.09	0.67
12mm diameter	m	0.05	0.40	0.34	0.11	0.85
16mm diameter	m	0.06	0.48	0.54	0.15	1.17
20mm diameter	m	0.07	0.56	0.86	0.21	1.63
25mm diameter	m	0.08	0.64	1.40	0.31	2.35

Steel fabric reinforcement laid in
concrete beds

A98 weighing 1.54kg per m2	m2	0.12	0.96	0.78	0.26	2.00
A142 weighing 2.22kg per m2	m2	0.14	1.12	0.92	0.31	2.35
A193 weighing 3.02kg per m2	m2	0.17	1.36	1.30	0.40	3.06
A252 weighing 3.95kg per m2	m2	0.19	1.52	1.60	0.47	3.59
B196 weighing 3.05kg per m2	m2	0.17	1.36	1.37	0.41	3.14
B283 weighing 3.73kg per m2	m2	0.18	1.44	1.55	0.45	3.44
B385 weighing 4.53kg per m2	m2	0.20	1.60	1.92	0.53	4.05
B503 weighing 5.93kg per m2	m2	0.22	1.76	2.48	0.64	4.88
B785 weighing 8.14kg per m2	m2	0.24	1.92	3.33	0.79	6.04
C283 weighing 2.61kg per m2	m2	0.15	1.20	1.28	0.37	2.85
C385 weighing 3.41kg per m2	m2	0.18	1.44	1.46	0.44	3.34
C503 weighing 4.34kg per m2	m2	0.20	1.60	1.88	0.52	4.00
C636 weighing 5.55kg per m2	m2	0.21	1.68	2.50	0.63	4.81
C785 weighing 6.72kg per m2	m2	0.23	1.84	2.95	0.72	5.51

	Unit	Hours	Hours £	Materials £	O & P £	Total £

Joints

Impregnated fibreboard expansion
joint, width

12.5mm thick

	Unit	Hours	Hours £	Materials £	O & P £	Total £
not exceeding 150mm thick	m	0.14	1.12	1.34	0.37	2.83
150 to 300mm thick	m	0.20	1.60	2.22	0.57	4.39
300 to 450mm thick	m	0.22	1.76	3.20	0.74	5.70
20mm thick						
not exceeding 150mm thick	m	0.16	1.28	1.96	0.49	3.73
150 to 300mm thick	m	0.22	1.76	2.95	0.71	5.42
300 to 450mm thick	m	0.24	1.92	4.41	0.95	7.28
25mm thick						
not exceeding 150mm thick	m	0.18	1.44	2.17	0.54	4.15
150 to 300mm thick	m	0.24	1.92	3.38	0.80	6.10
300 to 450mm thick	m	0.26	2.08	4.98	1.06	8.12

Concrete finishes

Treat surfaces of unset concrete

	Unit	Hours	Hours £	Materials £	O & P £	Total £
mechanical tamping	m2	0.07	0.56	0.00	0.08	0.64
power floating	m2	0.18	1.44	0.00	0.22	1.66
trowelling	m2	0.24	1.92	0.00	0.29	2.21
spade finish	m2	0.20	1.60	0.00	0.24	1.84
wood float finish	m2	0.16	1.28	0.00	0.19	1.47

Precast concrete

Reinforced concrete lintels
bedded in cement mortar (1:3)

	Unit	Hours	Hours £	Materials £	O & P £	Total £
100 x 75 x 900mm	nr	0.35	2.80	5.46	1.24	9.50
100 x 75 x 1000mm	nr	0.40	3.20	6.12	1.40	10.72
100 x 75 x 1200mm	nr	0.50	4.00	6.54	1.58	12.12
150 x 75 x 900mm	nr	0.45	3.60	5.88	1.42	10.90
150 x 75 x 1000mm	nr	0.45	3.60	6.65	1.54	11.79
150 x 75 x 1200mm	nr	0.55	4.40	7.98	1.86	14.24

	Unit	Hours	Hours £	Materials £	O & P £	Total £
Precast concrete copings bedded in cement mortar (1:3)						
75 x 150mm	m	0.75	6.00	5.87	1.78	13.65
75 x 300mm	m	0.90	7.20	11.50	2.81	21.51

	Unit	Hours	Hours £	Materials £	O & P £	Total £

BRICKWORK AND BLOCKWORK

Brickwork

Common bricks basic price £140
per thousand in cement mortar
(1:3)

walls						
half brick thick	m2	1.70	23.99	11.45	5.32	40.75
half brick thick built overhand	m2	2.20	31.04	11.45	6.37	48.87
half brick thick built against						
other work	m2	1.90	26.81	11.45	5.74	44.00
half brick thick built curved	m2	2.30	32.45	11.45	6.59	50.49
one brick thick	m2	2.80	39.51	22.90	9.36	71.77
one brick thick built curved	m2	3.40	47.97	22.90	10.63	81.51
one and a half brick thick	m2	3.50	49.39	34.35	12.56	96.30
two brick thick	m2	4.20	59.26	45.80	15.76	120.82
two brick thick built battered	m2	4.80	67.73	45.80	17.03	130.56
walls, facework one side						
half brick thick	m2	1.80	25.40	11.45	5.53	42.38
half brick thick built overhand	m2	2.30	32.45	11.45	6.59	50.49
half brick thick built against						
other work	m2	2.00	28.22	11.45	5.95	45.62
half brick thick built curved	m2	2.40	33.86	11.45	6.80	52.11
one brick thick	m2	2.90	40.92	22.90	9.57	73.39
one brick thick built curved	m2	3.50	49.39	22.90	10.84	83.13
one and a half brick thick	m2	3.60	50.80	34.35	12.77	97.92
two brick thick	m2	4.30	60.67	45.80	15.97	122.44
two brick thick built battered	m2	4.90	69.14	45.80	17.24	132.18
walls, facework both sides						
half brick thick	m2	1.90	26.81	11.45	5.74	44.00
half brick thick built overhand	m2	2.40	33.86	11.45	6.80	52.11
half brick thick built against						
other work	m2	2.10	29.63	11.45	6.16	47.24
half brick thick built curved	m2	2.50	35.28	11.45	7.01	53.73
one brick thick	m2	3.00	42.33	22.90	9.78	75.01

	Unit	Hours	Hours £	Materials £	O & P £	Total £
one brick thick built curved	m2	3.60	50.80	22.90	11.05	84.75
one and a half brick thick	m2	3.70	52.21	34.35	12.98	99.54
two brick thick	m2	4.40	62.08	45.80	16.18	124.07
two brick thick built battered	m2	5.00	70.55	45.80	17.45	133.80

Common bricks basic price £200
per thousand in cement mortar
(1:3)

walls

	Unit	Hours	Hours £	Materials £	O & P £	Total £
half brick thick	m2	1.70	23.99	16.60	6.09	46.68
half brick thick built overhand	m2	2.20	31.04	16.60	7.15	54.79
half brick thick built against other work	m2	1.90	26.81	16.60	6.51	49.92
half brick thick built curved	m2	2.30	32.45	16.60	7.36	56.41
one brick thick	m2	2.80	39.51	33.20	10.91	83.61
one brick thick built curved	m2	3.40	47.97	33.20	12.18	93.35
one and a half brick thick	m2	3.50	49.39	49.80	14.88	114.06
two brick thick	m2	4.20	59.26	86.40	21.85	167.51
two brick thick built battered	m2	4.80	67.73	45.80	17.03	130.56

walls, facework one side

	Unit	Hours	Hours £	Materials £	O & P £	Total £
half brick thick	m2	1.90	26.81	16.60	6.51	49.92
half brick thick built overhand	m2	2.40	33.86	16.60	7.57	58.03
half brick thick built against other work	m2	2.10	29.63	16.60	6.93	53.17
half brick thick built curved	m2	2.50	35.28	16.60	7.78	59.66
one brick thick	m2	3.00	42.33	33.20	11.33	86.86
one brick thick built curved	m2	3.60	50.80	33.20	12.60	96.60
one and a half brick thick	m2	3.70	52.21	49.80	15.30	117.31
two brick thick	m2	4.40	62.08	86.40	22.27	170.76
two brick thick built battered	m2	5.00	70.55	86.40	23.54	180.49

walls, facework both sides

	Unit	Hours	Hours £	Materials £	O & P £	Total £
half brick thick	m2	1.90	26.81	16.60	6.51	49.92
half brick thick built overhand	m2	2.40	33.86	16.60	7.57	58.03
half brick thick built against other work	m2	2.10	29.63	16.60	6.93	53.17
half brick thick built curved	m2	2.50	35.28	16.60	7.78	59.66
one brick thick	m2	3.00	42.33	33.20	11.33	86.86

	Unit	Hours	Hours £	Materials £	O & P £	Total £
one brick thick built curved	m2	3.60	50.80	33.20	12.60	96.60
one and a half brick thick	m2	3.70	52.21	49.80	15.30	117.31
two brick thick	m2	4.40	62.08	86.40	22.27	170.76
two brick thick built battered	m2	5.00	70.55	86.40	23.54	180.49

Common bricks basic price £140
per thousand in gauged mortar
(1:1:6)

	Unit	Hours	Hours £	Materials £	O & P £	Total £
walls						
half brick thick	m2	1.70	23.99	11.30	5.29	40.58
half brick thick built overhand	m2	2.20	31.04	11.30	6.35	48.69
half brick thick built against						
other work	m2	1.90	26.81	11.30	5.72	43.83
half brick thick built curved	m2	2.30	32.45	11.30	6.56	50.32
one brick thick	m2	2.80	39.51	22.60	9.32	71.42
one brick thick built curved	m2	3.40	47.97	22.60	10.59	81.16
one and a half brick thick	m2	3.50	49.39	33.90	12.49	95.78
two brick thick	m2	4.20	59.26	45.20	15.67	120.13
two brick thick built battered	m2	4.80	67.73	45.20	16.94	129.87
walls, facework one side						
half brick thick	m2	1.80	25.40	11.30	5.50	42.20
half brick thick built overhand	m2	2.30	32.45	11.30	6.56	50.32
half brick thick built against						
other work	m2	2.00	28.22	11.30	5.93	45.45
half brick thick built curved	m2	2.40	33.86	11.30	6.77	51.94
one brick thick	m2	2.90	40.92	22.60	9.53	73.05
one brick thick built curved	m2	3.50	49.39	22.60	10.80	82.78
one and a half brick thick	m2	3.60	50.80	33.90	12.70	97.40
two brick thick	m2	4.30	60.67	45.20	15.88	121.75
two brick thick built battered	m2	4.90	69.14	45.20	17.15	131.49
walls, facework both sides						
half brick thick	m2	1.90	26.81	11.30	5.72	43.83
half brick thick built overhand	m2	2.40	33.86	11.30	6.77	51.94
half brick thick built against						
other work	m2	2.10	29.63	11.30	6.14	47.07
half brick thick built curved	m2	2.50	35.28	11.30	6.99	53.56
one brick thick	m2	3.00	42.33	22.60	9.74	74.67

	Unit	Hours	Hours £	Materials £	O & P £	Total £
one brick thick built curved	m2	3.60	50.80	22.60	11.01	84.41
one and a half brick thick	m2	3.70	52.21	33.90	12.92	99.02
two brick thick	m2	4.40	62.08	45.20	16.09	123.38
two brick thick built battered	m2	5.00	70.55	45.20	17.36	133.11

Common bricks basic price £200
per thousand in gauged mortar
(1:3)

	Unit	Hours	Hours £	Materials £	O & P £	Total £
walls						
half brick thick	m2	1.70	23.99	16.45	6.07	46.50
half brick thick built overhand	m2	2.20	31.04	16.45	7.12	54.62
half brick thick built against						
other work	m2	1.90	26.81	16.45	6.49	49.75
half brick thick built curved	m2	2.30	32.45	16.45	7.34	56.24
one brick thick	m2	2.80	39.51	32.90	10.86	83.27
one brick thick built curved	m2	3.40	47.97	32.90	12.13	93.01
one and a half brick thick	m2	3.50	49.39	49.35	14.81	113.55
two brick thick	m2	4.20	59.26	65.80	18.76	143.82
two brick thick built battered	m2	4.80	67.73	65.80	20.03	153.56
walls, facework one side						
half brick thick	m2	1.80	25.40	16.45	6.28	48.13
half brick thick built overhand	m2	2.30	32.45	16.45	7.34	56.24
half brick thick built against						
other work	m2	2.00	28.22	16.45	6.70	51.37
half brick thick built curved	m2	2.40	33.86	16.45	7.55	57.86
one brick thick	m2	2.90	40.92	32.90	11.07	84.89
one brick thick built curved	m2	3.50	49.39	32.90	12.34	94.63
one and a half brick thick	m2	3.60	50.80	49.35	15.02	115.17
two brick thick	m2	4.30	60.67	65.80	18.97	145.44
two brick thick built battered	m2	4.90	69.14	65.80	20.24	155.18
walls, facework both sides						
half brick thick	m2	1.90	26.81	16.45	6.49	49.75
half brick thick built overhand	m2	2.40	33.86	16.45	7.55	57.86
half brick thick built against						
other work	m2	2.10	29.63	16.45	6.91	52.99
half brick thick built curved	m2	2.50	35.28	16.45	7.76	59.48

	Unit	Hours	Hours £	Materials £	O & P £	Total £
one brick thick	m2	3.00	42.33	32.90	11.28	86.51
one brick thick built curved	m2	3.60	50.80	32.90	12.55	96.25
one and a half brick thick	m2	3.70	52.21	49.35	15.23	116.79
two brick thick	m2	4.40	62.08	65.80	19.18	147.07
two brick thick built battered	m2	5.00	70.55	65.80	20.45	156.80

Class B engineering bricks basic
price £250 per thousand in
cement mortar (1:3)

walls

	Unit	Hours	Hours £	Materials £	O & P £	Total £
half brick thick	m2	1.80	25.40	20.45	6.88	52.73
half brick thick built overhand	m2	2.30	32.45	20.45	7.94	60.84
half brick thick built against other work	m2	2.00	28.22	20.45	7.30	55.97
half brick thick built curved	m2	2.40	33.86	20.45	8.15	62.46
one brick thick	m2	2.90	40.92	49.90	13.62	104.44
one brick thick built curved	m2	3.50	49.39	49.90	14.89	114.18
one and a half brick thick	m2	3.60	50.80	61.35	16.82	128.97
two brick thick	m2	4.30	60.67	81.90	21.39	163.96
two brick thick built battered	m2	4.90	69.14	81.90	22.66	173.69

walls, facework one side

	Unit	Hours	Hours £	Materials £	O & P £	Total £
half brick thick	m2	1.90	26.81	20.45	7.09	54.35
half brick thick built overhand	m2	2.40	33.86	20.45	8.15	62.46
half brick thick built against other work	m2	2.10	29.63	20.45	7.51	57.59
half brick thick built curved	m2	2.50	35.28	20.45	8.36	64.08
one brick thick	m2	3.00	42.33	49.90	13.83	106.06
one brick thick built curved	m2	3.60	50.80	49.90	15.10	115.80
one and a half brick thick	m2	3.70	52.21	61.35	17.03	130.59
two brick thick	m2	4.40	62.08	81.90	21.60	165.58
two brick thick built battered	m2	5.00	70.55	81.90	22.87	175.32

walls, facework both sides

	Unit	Hours	Hours £	Materials £	O & P £	Total £
half brick thick	m2	2.00	28.22	20.45	7.30	55.97
half brick thick built overhand	m2	2.50	35.28	20.45	8.36	64.08
half brick thick built against other work	m2	2.20	31.04	20.45	7.72	59.22
half brick thick built curved	m2	2.60	36.69	20.45	8.57	65.71

	Unit	Hours	Hours £	Materials £	O & P £	Total £
one brick thick	m2	3.10	43.74	49.90	14.05	107.69
one brick thick built curved	m2	3.70	52.21	49.90	15.32	117.42
one and a half brick thick	m2	3.80	53.62	61.35	17.25	132.21
two brick thick	m2	4.50	63.50	81.90	21.81	167.20
two brick thick built battered	m2	5.60	79.02	81.90	24.14	185.05

Class B engineering bricks basic
price £350 per thousand in
cement mortar (1:3)

walls

	Unit	Hours	Hours £	Materials £	O & P £	Total £
half brick thick	m2	1.80	25.40	28.63	8.10	62.13
half brick thick built overhand	m2	2.30	32.45	28.63	9.16	70.25
half brick thick built against other work	m2	2.00	28.22	28.63	8.53	65.38
half brick thick built curved	m2	2.40	33.86	28.63	9.37	71.87
one brick thick	m2	2.90	40.92	57.26	14.73	112.91
one brick thick built curved	m2	3.50	49.39	57.26	16.00	122.64
one and a half brick thick	m2	3.60	50.80	85.89	20.50	157.19
two brick thick	m2	4.30	60.67	114.57	26.29	201.53
two brick thick built battered	m2	4.90	69.14	114.57	27.56	211.27

walls, facework one side

	Unit	Hours	Hours £	Materials £	O & P £	Total £
half brick thick	m2	1.90	26.81	28.63	8.32	63.75
half brick thick built overhand	m2	2.40	33.86	28.63	9.37	71.87
half brick thick built against other work	m2	2.10	29.63	28.63	8.74	67.00
half brick thick built curved	m2	2.50	35.28	28.63	9.59	73.49
one brick thick	m2	3.00	42.33	57.26	14.94	114.53
one brick thick built curved	m2	3.60	50.80	57.26	16.21	124.26
one and a half brick thick	m2	3.70	52.21	85.89	20.71	158.81
two brick thick	m2	4.40	62.08	114.57	26.50	203.15
two brick thick built battered	m2	5.00	70.55	114.57	27.77	212.89

walls, facework both sides

	Unit	Hours	Hours £	Materials £	O & P £	Total £
half brick thick	m2	2.00	28.22	28.63	8.53	65.38
half brick thick built overhand	m2	2.50	35.28	28.63	9.59	73.49
half brick thick built against other work	m2	2.20	31.04	28.63	8.95	68.62
half brick thick built curved	m2	2.60	36.69	28.63	9.80	75.11

	Unit	Hours	Hours £	Materials £	O & P £	Total £
one brick thick	m2	3.10	43.74	57.26	15.15	116.15
one brick thick built curved	m2	3.70	52.21	57.26	16.42	125.89
one and a half brick thick	m2	3.80	53.62	85.89	20.93	160.43
two brick thick	m2	4.50	63.50	114.57	26.71	204.77
two brick thick built battered	m2	5.60	79.02	114.57	29.04	222.62

Facing bricks basic price £250
per thousand in gauged mortar (1:3)

	Unit	Hours	Hours £	Materials £	O & P £	Total £
walls						
half brick thick	m2	1.70	23.99	20.30	6.64	50.93
half brick thick built overhand	m2	2.20	31.04	20.30	7.70	59.04
half brick thick built against						
other work	m2	1.90	26.81	20.30	7.07	54.18
half brick thick built curved	m2	2.30	32.45	20.30	7.91	60.67
one brick thick	m2	2.80	39.51	40.60	12.02	92.12
one brick thick built curved	m2	3.40	47.97	40.60	13.29	101.86

Extra over for fair face and flush
pointing

	Unit	Hours	Hours £	Materials £	O & P £	Total £
stretcher bond	m2	0.60	8.47	0.10	1.28	9.85
flemish bond	m2	0.70	9.88	0.10	1.50	11.47
margins	m	0.10	1.41	0.10	0.23	1.74
soffits of flat arches	m	0.65	9.17	0.10	1.39	10.66
soffits of curved arches	m	1.20	16.93	0.10	2.55	19.59

Isolated casings

	Unit	Hours	Hours £	Materials £	O & P £	Total £
half brick thick, facework all						
round, stretcher bond	m2	2.30	32.45	20.40	7.93	60.78
half brick thick, facework all						
round, flemish bond	m2	2.40	33.86	20.40	8.14	62.40

Isolated piers

	Unit	Hours	Hours £	Materials £	O & P £	Total £
one brick thick, facework all						
round, stretcher bond	m2	3.40	47.97	4.08	7.81	59.86
one brick thick, facework all						
round, flemish bond	m2	3.50	49.39	40.80	13.53	103.71

	Unit	Hours	Hours £	Materials £	O & P £	Total £
Arches						
215mm high, 102mm wide, flat	m	1.38	19.47	4.36	3.57	27.41
215mm high, 215mm wide, flat	m	2.05	28.93	7.34	5.44	41.71
215mm high, 102mm wide, segmental	m	2.76	38.94	8.27	7.08	54.30
215mm high, 215mm wide, segmental	m	3.72	52.49	11.56	9.61	73.66
Projections						
112mm high, 225mm wide, stretcher bond	m	0.48	6.77	5.14	1.79	13.70
112mm high, 225mm wide, flemish bond	m	0.60	8.47	5.14	2.04	15.65
225mm high, 225mm wide, flemish bond	m	0.85	11.99	10.28	3.34	25.61
112mm high, 328mm wide, stretcher bond	m	0.75	10.58	7.17	2.66	20.42
112mm high, 328mm wide, flemish bond	m	0.95	13.40	7.17	3.09	23.66
Facing bricks basic price £400 per thousand in gauged mortar (1:3)						
walls						
half brick thick	m2	1.70	23.99	28.55	7.88	60.42
half brick thick built overhand	m2	2.20	31.04	28.55	8.94	68.53
half brick thick built against other work	m2	1.90	26.81	28.55	8.30	63.66
half brick thick built curved	m2	2.30	32.45	28.55	9.15	70.15
one brick thick	m2	2.80	39.51	57.10	14.49	111.10
one brick thick built curved	m2	3.40	47.97	57.10	15.76	120.84
Extra over for fair face and flush pointing						
stretcher bond	m2	0.60	8.47	0.10	1.28	9.85
flemish bond	m2	0.70	9.88	0.10	1.50	11.47
margins	m	0.10	1.41	0.10	0.23	1.74

	Unit	Hours	Hours £	Materials £	O & P £	Total £
soffits of flat arches	m	0.65	9.17	0.10	1.39	10.66
soffits of curved arches	m	1.20	16.93	0.10	2.55	19.59
Isolated casings						
half brick thick, facework all round, stretcher bond	m2	2.30	32.45	28.65	9.17	70.27
half brick thick, facework all round, flemish bond	m2	2.40	33.86	28.65	9.38	71.89
Isolated piers						
one brick thick, facework all round, stretcher bond	m2	3.40	47.97	57.30	15.79	121.07
one brick thick, facework all round, flemish bond	m2	3.50	49.39	57.30	16.00	122.69
Arches						
215mm high, 102mm wide, flat	m	1.38	19.47	6.98	3.97	30.42
215mm high, 215mm wide, flat	m	2.05	28.93	11.74	6.10	46.77
215mm high, 102mm wide, segmental	m	2.76	38.94	13.23	7.83	60.00
215mm high, 215mm wide, segmental	m	3.72	52.49	18.50	10.65	81.64
Projections						
112mm high, 225mm wide, stretcher bond	m	0.48	6.77	8.22	2.25	17.24
112mm high, 225mm wide, flemish bond	m	0.60	8.47	8.22	2.50	19.19
225mm high, 225mm wide, flemish bond	m	0.85	11.99	16.38	4.26	32.63
112mm high, 328mm wide, stretcher bond	m	0.75	10.58	11.47	3.31	25.36
112mm high, 328mm wide, flemish bond	m	0.95	13.40	11.47	3.73	28.61

	Unit	Hours	Hours £	Materials £	O & P £	Total £
Facing bricks basic price £500 per thousand in gauged mortar (1:3)						
walls						
half brick thick	m2	1.70	23.99	34.85	8.83	67.66
half brick thick built overhand	m2	2.20	31.04	34.85	9.88	75.78
half brick thick built against other work	m2	1.90	26.81	34.85	9.25	70.91
half brick thick built curved	m2	2.30	32.45	34.85	10.10	77.40
one brick thick	m2	2.80	39.51	69.70	16.38	125.59
one brick thick built curved	m2	3.40	47.97	69.70	17.65	135.33
Extra over for fair face and flush pointing						
stretcher bond	m2	0.60	8.47	0.10	1.28	9.85
flemish bond	m2	0.70	9.88	0.10	1.50	11.47
margins	m	0.10	1.41	0.10	0.23	1.74
soffits of flat arches	m	0.65	9.17	0.10	1.39	10.66
soffits of curved arches	m	1.20	16.93	0.10	2.55	19.59
Isolated casings						
half brick thick, facework all round, stretcher bond	m2	2.30	32.45	34.85	10.10	77.40
half brick thick, facework all round, flemish bond	m2	2.40	33.86	34.85	10.31	79.02
Isolated piers						
one brick thick, facework all round, stretcher bond	m2	3.40	47.97	69.70	17.65	135.33
one brick thick, facework all round, flemish bond	m2	3.50	49.39	69.70	17.86	136.95
Arches						
215mm high, 102mm wide, flat	m	1.38	19.47	8.73	4.23	32.43
215mm high, 215mm wide, flat	m	2.05	28.93	14.68	6.54	50.15

	Unit	Hours	Hours £	Materials £	O & P £	Total £
215mm high, 102mm wide, segmental	m	2.76	38.94	16.54	8.32	63.81
215mm high, 215mm wide, segmental	m	3.72	52.49	23.13	11.34	86.96

Projections

112mm high, 225mm wide, stretcher bond	m	0.48	6.77	10.28	2.56	19.61
112mm high, 225mm wide, flemish bond	m	0.60	8.47	10.28	2.81	21.56
225mm high, 225mm wide, flemish bond	m	0.85	11.99	16.38	4.26	32.63
112mm high, 328mm wide, stretcher bond	m	0.75	10.58	20.48	4.66	35.72
112mm high, 328mm wide, flemish bond	m	0.95	13.40	14.34	4.16	31.91

Blockwork

Concrete dense aggregate blocks
in gauged mortar (1:1:6)

walls and partitions						
75mm thick, solid	m2	0.98	13.83	6.93	3.11	23.87
100mm thick, solid	m2	1.14	16.09	7.08	3.47	26.64
140mm thick, solid	m2	1.32	18.63	14.05	4.90	37.58
140mm thick, hollow	m2	1.43	20.18	14.29	5.17	39.64
190mm thick, hollow	m2	1.76	24.83	15.97	6.12	46.92
215mm thick, hollow	m2	2.02	28.50	16.15	6.70	51.35
skins of hollow walls						
75mm thick, solid	m2	1.12	15.80	6.93	3.41	26.14
100mm thick, solid	m2	1.24	17.50	7.08	3.69	28.26
140mm thick, solid	m2	1.27	17.92	14.29	4.83	37.04
215mm thick, hollow	m2	2.15	30.34	16.15	6.97	53.46

	Unit	Hours	Hours £	Materials £	O & P £	Total £
piers and chimney stacks						
75mm thick, solid	m2	1.28	18.06	6.93	3.75	28.74
100mm thick, solid	m2	1.40	19.75	7.08	4.03	30.86
140mm thick, solid	m2	1.68	23.70	14.29	5.70	43.69
isolated casings						
75mm thick, solid	m2	1.42	20.04	6.93	4.04	31.01
100mm thick, solid	m2	1.90	26.81	7.08	5.08	38.97
140mm thick, solid	m2	2.15	30.34	14.29	6.69	51.32
Extra over for fair face and flush pointing						
one side	m2	0.16	2.26	0.00	0.34	2.60
two sides	m2	0.33	4.66	0.00	0.70	5.35
Bonding ends of blockwork to brickwork						
75mm thick, solid	m2	0.30	4.23	0.00	0.63	4.87
100mm thick, solid	m2	0.38	5.36	0.00	0.80	6.17
140mm thick, solid	m2	0.44	6.21	0.00	0.93	7.14
140mm thick, hollow	m2	0.48	6.77	0.00	1.02	7.79
190mm thick, hollow	m2	0.55	7.76	0.00	1.16	8.92
215mm thick, hollow	m2	0.62	8.75	0.00	1.31	10.06
Close cavities at ends of walls and jambs with 100mm thick blockwork and bitumen damp proof course						
bedded in gauged mortar (1:1:6)	m	0.20	2.82	1.42	0.64	4.88
Close cavities at tops of walls with blocks laid flat in gauged mortar (1:1:6)						
100mm thick, solid	m	0.24	3.39	2.84	0.93	7.16
140mm thick, solid	m	0.28	3.95	3.12	1.06	8.13

	Unit	Hours	Hours £	Materials £	O & P £	Total £
Lightweight aerated concrete blocks in gauged mortar (1:1:6)						
walls and partitions						
75mm thick	m2	0.90	12.70	6.84	2.93	22.47
100mm thick	m2	1.05	14.82	7.95	3.41	26.18
125mm thick	m2	1.20	16.93	9.14	3.91	29.98
140mm thick	m2	1.26	17.78	13.62	4.71	36.11
150mm thick	m2	1.34	18.91	15.27	5.13	39.30
190mm thick	m2	1.60	22.58	18.32	6.13	47.03
215mm thick	m2	2.00	28.22	21.12	7.40	56.74
255mm thick	m2	2.10	29.63	24.87	8.18	62.68
skins of hollow walls						
75mm thick	m2	1.00	14.11	6.84	3.14	24.09
100mm thick	m2	1.15	16.23	7.95	3.63	27.80
125mm thick	m2	1.30	18.34	9.14	4.12	31.61
150mm thick	m2	1.44	20.32	15.27	5.34	40.93
piers and chimney stacks						
140mm thick	m2	1.44	20.32	13.62	5.09	39.03
190mm thick	m2	1.52	21.45	18.32	5.97	45.73
215mm thick	m2	1.62	22.86	21.12	6.60	50.57
isolated casings						
75mm thick	m2	1.60	22.58	13.62	5.43	41.63
100mm thick	m2	1.92	27.09	18.32	6.81	52.22
140mm thick	m2	2.07	29.21	21.12	7.55	57.88
Extra over for fair face and flush pointing						
one side	m2	0.16	2.26	0.00	0.34	2.60
two sides	m2	0.33	4.66	0.00	0.70	5.35

	Unit	Hours	Hours £	Materials £	O & P £	Total £
Bonding ends of blockwork to brickwork						
75mm thick	m	0.25	3.53	0.00	0.53	4.06
100mm thick	m	0.32	4.52	0.00	0.68	5.19
125mm thick	m	0.36	5.08	0.00	0.76	5.84
140mm thick	m	0.40	5.64	0.00	0.85	6.49
150mm thick	m	0.44	6.21	0.00	0.93	7.14
190mm thick	m	0.50	7.06	0.00	1.06	8.11
215mm thick	m	0.55	7.76	0.00	1.16	8.92
255mm thick	m	0.58	8.18	0.00	1.23	9.41
Close cavities at ends of walls and jambs with 100mm thick blockwork and bitumen damp proof course bedded in gauged mortar (1:1:6)	m	0.20	2.82	1.54	0.65	5.02
Close cavities at tops of walls with blocks laid flat in gauged mortar (1:1:6)						
100mm thick, solid	m	0.95	13.40	14.34	4.16	31.91
140mm thick, solid	m	0.95	13.40	14.34	4.16	31.91
Lightweight aerated concrete blocks, smooth faced, in gauged mortar (1:1:6)						
walls and partitions						
100mm thick	m2	1.15	16.23	15.82	4.81	36.85
140mm thick	m2	1.36	19.19	22.20	6.21	47.60
190mm thick	m2	1.70	23.99	29.48	8.02	61.49
215mm thick	m2	2.10	29.63	33.87	9.53	73.03
piers and chimney stacks						
140mm thick	m2	1.54	21.73	22.20	6.59	50.52
190mm thick	m2	1.62	22.86	29.48	7.85	60.19
215mm thick	m2	1.72	24.27	33.87	8.72	66.86

	Unit	Hours	Hours £	Materials £	O & P £	Total £
isolated casings						
100mm thick	m2	2.02	28.50	15.82	6.65	50.97
140mm thick	m2	2.16	30.48	22.20	7.90	60.58

Extra over for fair face and flush
pointing

one side	m2	0.16	2.26	0.00	0.34	2.60
two sides	m2	0.33	4.66	0.00	0.70	5.35

Damp proof courses

Hessian based bitumen damp
proof course in gauged mortar
(1:1:6)

horizontal, width						
over 225mm	m2	0.35	4.94	8.14	1.96	15.04
112mm	m	0.05	0.71	0.95	0.25	1.90
vertical, width						
over 225mm	m2	0.40	5.64	8.14	2.07	15.85
112mm	m	0.07	0.99	0.95	0.29	2.23

Fibre based damp proof course
in gauged mortar (1:1:6)

horizontal, width						
over 225mm	m2	0.40	5.64	5.68	1.70	13.02
112mm	m	0.05	0.71	0.67	0.21	1.58
vertical, width						
over 225mm	m2	0.45	6.35	5.68	1.80	13.83
112mm	m	0.07	0.99	0.67	0.25	1.91

	Unit	Hours	Hours £	Materials £	O & P £	Total £
Pitch polymer damp proof course in gauged mortar (1:1:6)						
horizontal, width						
over 225mm	m2	0.40	5.64	6.81	1.87	14.32
112mm	m	0.05	0.71	0.81	0.23	1.74
vertical, width						
over 225mm	m2	0.45	6.35	6.81	1.97	15.13
112mm	m	0.07	0.99	0.81	0.27	2.07
Bitumen damp proof course with lead core in gauged mortar (1:1:6)						
horizontal, width						
over 225mm	m2	0.50	7.06	17.25	3.65	27.95
112mm	m	0.06	0.85	2.12	0.44	3.41
vertical, width						
over 225mm	m2	0.55	7.76	17.25	3.75	28.76
112mm	m	0.08	1.13	2.12	0.49	3.74
Two courses of slates bedded in cement mortar (1:3)						
horizontal, width						
over 225mm	m2	0.87	12.28	32.10	6.66	51.03
112mm	m	0.33	4.66	3.10	1.16	8.92
vertical, width						
over 225mm	m2	0.95	13.40	32.10	6.83	52.33
112mm	m	0.35	4.94	3.10	1.21	9.24
Three courses of waterproofing liquid brush on concrete surfaces						
vertically	m2	0.45	6.35	7.12	2.02	15.49
horizontally	m2	0.50	7.06	7.12	2.13	16.30

	Unit	Hours	Hours £	Materials £	O & P £	Total £

Sundries

Expanded polystyrene cavity wall
insulation sheeting, thickness

	Unit	Hours	Hours £	Materials £	O & P £	Total £
25mm	m2	0.20	2.82	4.88	1.16	8.86
50mm	m2	0.22	3.10	6.58	1.45	11.14

Galvanised steel brick
reinforcement, width

65mm	m	0.07	0.99	0.35	0.20	1.54
115mm	m	0.09	1.27	0.42	0.25	1.94
175mm	m	0.12	1.69	0.68	0.36	2.73
225mm	m	0.15	2.12	0.98	0.46	3.56

Point frames with mastic

one side	m	0.14	1.98	0.51	0.37	2.86
both sides	m	0.22	3.10	1.02	0.62	4.74

Point frames with polysulphide
sealant

one side	m	0.14	1.98	1.67	0.55	4.19
both sides	m	0.22	3.10	3.34	0.97	7.41

Rake out joints for flashings and
point up on completion

horizontal	m	0.09	1.27	0.20	0.22	1.69
stepped	m	0.18	2.54	0.24	0.42	3.20

	Unit	Hours	Hours £	Materials £	O & P £	Total £

MASONRY

Walling

Random rubble walling, laid dry, thickness

300mm	m2	3.00	30.33	42.85	10.98	84.16
450mm	m2	3.50	35.39	64.23	14.94	114.56
500mm	m2	3.75	37.91	72.45	16.55	126.92

Random rubble walling, laid dry, battered one side, thickness

300mm	m2	3.25	32.86	42.85	11.36	87.06
450mm	m2	3.75	37.91	64.23	15.32	117.46
500mm	m2	4.00	40.44	72.45	16.93	129.82

Random rubble walling, laid dry, battered both sides, thickness

300mm	m2	3.50	35.39	42.85	11.74	89.97
450mm	m2	4.00	40.44	64.23	15.70	120.37
500mm	m2	4.25	42.97	72.45	17.31	132.73

Random rubble walling, laid in gauged mortar (1:1:6), thickness

300mm	m2	3.20	32.35	46.05	11.76	90.16
450mm	m2	3.70	37.41	69.03	15.97	122.40
500mm	m2	3.90	39.43	77.56	17.55	134.54

Random rubble walling, laid in gauged mortar (1:1:6), battered one side, thickness

300mm	m2	3.45	34.88	46.05	12.14	93.07
450mm	m2	4.00	40.44	69.03	16.42	125.89
500mm	m2	4.15	41.96	77.56	17.93	137.44

	Unit	Hours	Hours £	Materials £	O & P £	Total £
gauged mortar (1:1:6), battered both sides, thickness						
300mm	m2	3.95	39.93	46.05	12.90	98.88
450mm	m2	4.50	45.50	69.03	17.18	131.70
500mm	m2	4.65	47.01	77.56	18.69	143.26
Irregular coursed rubble walling, laid in gauged mortar (1:1:6), thickness						
300mm	m2	2.10	21.23	46.05	10.09	77.37
450mm	m2	2.90	29.32	69.03	14.75	113.10
500mm	m2	3.40	34.37	77.56	16.79	128.72
Coursed rubble walling, laid in gauged mortar (1:1:6), thickness						
300mm	m2	2.40	24.26	46.05	10.55	80.86
450mm	m2	3.20	32.35	69.03	15.21	116.59
500mm	m2	3.70	37.41	77.56	17.25	132.21
Fair raking cutting on stone walling, thickness						
300mm	m	0.80	8.09	2.45	1.58	12.12
450mm	m	1.20	12.13	3.97	2.42	18.52
500mm	m	1.80	18.20	4.56	3.41	26.17
Form level bed on stone walling, thickness						
300mm	m	0.20	2.02	0.70	0.41	3.13
450mm	m	0.35	3.54	1.12	0.70	5.36
500mm	m	0.50	5.06	1.23	0.94	7.23

	Unit	Hours	Hours £	Materials £	O & P £	Total £

Sundries

Form holes for pipes up to 50mm diameter through stone walling, thickness

	Unit	Hours	Hours £	Materials £	O & P £	Total £
300mm	m	1.20	12.13	0.00	1.82	13.95
450mm	m	1.80	18.20	0.00	2.73	20.93
500mm	m	2.65	26.79	0.00	4.02	30.81

Form holes for pipes 50-100mm diameter through stone walling, thickness

	Unit	Hours	Hours £	Materials £	O & P £	Total £
300mm	m	1.40	14.15	0.00	2.12	16.28
450mm	m	2.00	20.22	0.00	3.03	23.25
500mm	m	2.85	28.81	0.00	4.32	33.14

Form holes for pipes over 100mm diameter through stone walling, thickness

	Unit	Hours	Hours £	Materials £	O & P £	Total £
300mm	m	2.40	24.26	0.00	3.64	27.90
450mm	m	3.00	30.33	0.00	4.55	34.88
500mm	m	3.85	38.92	0.00	5.84	44.76

Build in ends of steel sections in stone walling, size

	Unit	Hours	Hours £	Materials £	O & P £	Total £
not exceeding 250mm deep	m	0.70	7.08	0.00	1.06	8.14
250-500mm deep	m	0.90	9.10	0.00	1.36	10.46
over 500mm deep	m	1.10	11.12	0.00	1.67	12.79

Mortices in stone walling, size

	Unit	Hours	Hours £	Materials £	O & P £	Total £
50 x 50 x 100mm	nr	0.60	6.07	0.00	0.91	6.98
50 x 50 x 150mm	nr	0.65	6.57	0.00	0.99	7.56
75 x 75 x 100mm	nr	0.70	7.08	0.00	1.06	8.14
75 x 75 x 150mm	nr	0.75	7.58	0.00	1.14	8.72

	Unit	Hours	Hours £	Materials £	O & P £	Total £
Grout up mortices in cement mortar (1:3)						
50 x 50 x 100mm	nr	0.10	1.01	0.10	0.17	1.28
50 x 50 x 150mm	nr	0.15	1.52	0.15	0.25	1.92
75 x 75 x 100mm	nr	0.15	1.52	0.25	0.26	2.03
75 x 75 x 150mm	nr	0.20	2.02	0.30	0.35	2.67

	Unit	Hours	Hours £	Materials £	O & P £	Total £

CARPENTRY AND JOINERY

Sawn softwood, untreated

Floors

50 x 100mm	m	0.16	1.66	1.14	0.42	3.22
50 x 125mm	m	0.17	1.77	1.40	0.48	3.64
50 x 150mm	m	0.18	1.87	1.72	0.54	4.13
75 x 125mm	m	0.17	1.77	2.42	0.63	4.82
75 x 150mm	m	0.18	1.87	2.82	0.70	5.40
75 x 200mm	m	0.19	1.98	3.30	0.79	6.07
75 x 225mm	m	0.20	2.08	4.12	0.93	7.13

Partitions

38 x 75mm	m	0.15	1.56	1.02	0.39	2.97
38 x 100mm	m	0.16	1.66	1.12	0.42	3.20
50 x 75mm	m	0.16	1.66	1.10	0.41	3.18
50 x 100mm	m	0.17	1.77	1.14	0.44	3.34

Flat roofs

38 x 100mm	m	0.10	1.04	1.12	0.32	2.48
50 x 75mm	m	0.11	1.14	1.10	0.34	2.58
50 x 100mm	m	0.12	1.25	1.14	0.36	2.75
50 x 125mm	m	0.13	1.35	1.40	0.41	3.16
50 x 150mm	m	0.15	1.56	1.72	0.49	3.77
75 x 125mm	m	0.16	1.66	2.42	0.61	4.70

Pitched roofs

38 x 100mm	m	0.14	1.46	1.12	0.39	2.96
50 x 75mm	m	0.15	1.56	1.10	0.40	3.06
50 x 100mm	m	0.16	1.66	1.14	0.42	3.22
50 x 125mm	m	0.17	1.77	1.40	0.48	3.64
50 x 150mm	m	0.19	1.98	1.72	0.55	4.25
75 x 125mm	m	0.20	2.08	2.42	0.68	5.18
75 x 150mm	m	0.23	2.40	2.80	0.78	5.98

	Unit	Hours	Hours £	Materials £	O & P £	Total £
Kerbs and bearers						
25 x 75mm	m	0.13	1.35	0.62	0.30	2.27
25 x 100mm	m	0.14	1.46	0.70	0.32	2.48
25 x 150mm	m	0.17	1.77	0.98	0.41	3.16
38 x 75mm	m	0.15	1.56	0.81	0.36	2.73
38 x 100mm	m	0.16	1.66	1.12	0.42	3.20
38 x 150mm	m	0.18	1.87	1.18	0.46	3.51
50 x 50mm	m	0.16	1.66	0.86	0.38	2.90
50 x 75mm	m	0.17	1.77	1.10	0.43	3.30
50 x 100mm	m	0.18	1.87	1.14	0.45	3.46
75 x 75mm	m	0.20	2.08	1.22	0.50	3.80
75 x 100mm	m	0.22	2.29	1.66	0.59	4.54
75 x 125mm	m	0.24	2.50	2.42	0.74	5.65
Solid strutting						
38 x 100mm	m	0.40	4.16	1.12	0.79	6.07
50 x 100mm	m	0.44	4.58	1.14	0.86	6.57
50 x 125mm	m	0.48	4.99	1.28	0.94	7.21
50 x 150mm	m	0.52	5.41	1.40	1.02	7.83
Herringbone strutting, 50 x 50mm to joists, depth						
125mm	m	0.60	6.24	1.20	1.12	8.56
150mm	m	0.60	6.24	1.24	1.12	8.60
175mm	m	0.60	6.24	1.28	1.13	8.65
240mm	m	0.60	6.24	1.32	1.13	8.69
Sawn softwood, impregnated						
Floors						
50 x 100mm	m	0.16	1.66	1.20	0.43	3.29
50 x 125mm	m	0.17	1.77	1.46	0.48	3.71
50 x 150mm	m	0.18	1.87	1.80	0.55	4.22
75 x 125mm	m	0.17	1.77	2.54	0.65	4.95
75 x 150mm	m	0.18	1.87	2.94	0.72	5.53

	Unit	Hours	Hours £	Materials £	O & P £	Total £
75 x 200mm	m	0.19	1.98	3.34	0.80	6.11
75 x 225mm	m	0.20	2.08	4.22	0.95	7.25

Partitions

	Unit	Hours	Hours £	Materials £	O & P £	Total £
38 x 75mm	m	0.15	1.56	1.06	0.39	3.01
38 x 100mm	m	0.16	1.66	1.15	0.42	3.24
50 x 75mm	m	0.16	1.66	1.16	0.42	3.25
50 x 100mm	m	0.17	1.77	1.20	0.45	3.41

Flat roofs

	Unit	Hours	Hours £	Materials £	O & P £	Total £
38 x 100mm	m	0.10	1.04	1.15	0.33	2.52
50 x 75mm	m	0.11	1.14	1.16	0.35	2.65
50 x 100mm	m	0.12	1.25	1.20	0.37	2.82
50 x 125mm	m	0.13	1.35	1.46	0.42	3.23
50 x 150mm	m	0.15	1.56	1.80	0.50	3.86
75 x 125mm	m	0.16	1.66	2.54	0.63	4.83

Pitched roofs

	Unit	Hours	Hours £	Materials £	O & P £	Total £
38 x 100mm	m	0.14	1.46	1.15	0.39	3.00
50 x 75mm	m	0.15	1.56	1.16	0.41	3.13
50 x 100mm	m	0.16	1.66	1.20	0.43	3.29
50 x 125mm	m	0.17	1.77	1.46	0.48	3.71
50 x 150mm	m	0.19	1.98	1.80	0.57	4.34
75 x 125mm	m	0.20	2.08	2.54	0.69	5.31
75 x 150mm	m	0.23	2.40	2.94	0.80	6.14

Kerbs and bearers

	Unit	Hours	Hours £	Materials £	O & P £	Total £
25 x 75mm	m	0.13	1.35	0.66	0.30	2.31
25 x 100mm	m	0.14	1.46	0.75	0.33	2.54
25 x 150mm	m	0.17	1.77	1.05	0.42	3.24
38 x 75mm	m	0.15	1.56	0.86	0.36	2.78
38 x 100mm	m	0.16	1.66	1.15	0.42	3.24
38 x 150mm	m	0.18	1.87	1.22	0.46	3.56
50 x 50mm	m	0.16	1.66	0.90	0.38	2.95
50 x 75mm	m	0.17	1.77	1.16	0.44	3.37
50 x 100mm	m	0.18	1.87	1.20	0.46	3.53

	Unit	Hours	Hours £	Materials £	O & P £	Total £
75 x 75mm	m	0.20	2.08	1.27	0.50	3.85
75 x 100mm	m	0.22	2.29	1.72	0.60	4.61
75 x 125mm	m	0.24	2.50	2.54	0.76	5.79

Solid strutting

	Unit	Hours	Hours £	Materials £	O & P £	Total £
38 x 100mm	m	0.40	4.16	1.16	0.80	6.12
50 x 100mm	m	0.44	4.58	1.18	0.86	6.62
50 x 125mm	m	0.48	4.99	1.33	0.95	7.27
50 x 150mm	m	0.52	5.41	1.45	1.03	7.89

Herringbone strutting, 50 x 50mm
to joists, depth

	Unit	Hours	Hours £	Materials £	O & P £	Total £
125mm	m	0.60	6.24	1.24	1.12	8.60
150mm	m	0.60	6.24	1.29	1.13	8.66
175mm	m	0.60	6.24	1.33	1.14	8.71
240mm	m	0.60	6.24	1.36	1.14	8.74

Gutters and fascias

Marine quality plywood in gutters
over 300mm wide, thickness

	Unit	Hours	Hours £	Materials £	O & P £	Total £
12mm	m2	0.90	9.36	10.94	3.05	23.35
18mm	m2	1.00	10.40	15.03	3.81	29.24
25mm	m2	1.10	11.44	17.45	4.33	33.22

Marine quality plywood in gutters
150mm wide, thickness

	Unit	Hours	Hours £	Materials £	O & P £	Total £
12mm	m	0.28	2.91	1.82	0.71	5.44
18mm	m	0.32	3.33	2.51	0.88	6.71
25mm	m	0.36	3.74	2.91	1.00	7.65

	Unit	Hours	Hours £	Materials £	O & P £	Total £
Marine quality plywood in gutters 300mm wide, thickness						
12mm	m	0.36	3.74	3.42	1.07	8.24
18mm	m	0.40	4.16	4.96	1.37	10.49
25mm	m	0.44	4.58	5.67	1.54	11.78
Marine quality plywood in eaves, verges, soffits, fascias over 300mm wide, thickness						
12mm	m2	0.80	8.32	10.94	2.89	22.15
18mm	m2	0.90	9.36	15.03	3.66	28.05
25mm	m2	1.00	10.40	17.45	4.18	32.03
Marine quality plywood in gutters 150mm wide, thickness						
12mm	m	0.24	2.50	1.82	0.65	4.96
18mm	m	0.28	2.91	2.51	0.81	6.24
25mm	m	0.32	3.33	2.91	0.94	7.17
Marine quality plywood in gutters 300mm wide, thickness						
12mm	m	0.32	3.33	3.42	1.01	7.76
18mm	m	0.36	3.74	4.96	1.31	10.01
25mm	m	0.40	4.16	5.67	1.47	11.30
Wrought softwood supports						
Bearers, framed						
25 x 50mm	m	0.09	0.94	0.38	0.20	1.51
38 x 50mm	m	0.11	1.14	0.60	0.26	2.01
50 x 50mm	m	0.12	1.25	0.70	0.29	2.24
50 x 75mm	m	0.14	1.46	0.78	0.34	2.57
75 x 75mm	m	0.16	1.66	0.97	0.40	3.03

	Unit	Hours	Hours £	Materials £	O & P £	Total £
Bearers, plugged and screwed						
25 x 50mm	m	0.14	1.46	0.48	0.29	2.23
38 x 50mm	m	0.16	1.66	0.70	0.35	2.72
50 x 50mm	m	0.17	1.77	0.80	0.39	2.95
50 x 75mm	m	0.19	1.98	0.88	0.43	3.28
75 x 75mm	m	0.21	2.18	1.07	0.49	3.74

Metal fixings

Galvanised mild steel joist hangars,
built in

	Unit	Hours	Hours £	Materials £	O & P £	Total £
38 x 100mm	nr	0.10	1.04	2.05	0.46	3.55
38 x 125mm	nr	0.10	1.04	2.10	0.47	3.61
38 x 150mm	nr	0.10	1.04	2.18	0.48	3.70
38 x 175mm	nr	0.10	1.04	2.20	0.49	3.73
50 x 100mm	nr	0.12	1.25	2.05	0.49	3.79
50 x 125mm	nr	0.12	1.25	2.10	0.50	3.85
50 x 150mm	nr	0.12	1.25	2.18	0.51	3.94
50 x 175mm	nr	0.12	1.25	2.20	0.52	3.97
50 x 200mm	nr	0.12	1.25	2.25	0.52	4.02
50 x 225mm	nr	0.12	1.25	2.30	0.53	4.08
75 x 150mm	nr	0.14	1.46	2.46	0.59	4.50
75 x 175mm	nr	0.14	1.46	2.50	0.59	4.55
75 x 200mm	nr	0.14	1.46	2.74	0.63	4.83
75 x 225mm	nr	0.14	1.46	2.78	0.64	4.87

Galvanised mild steel square
toothed timber connectors, single
sided, diameter

	Unit	Hours	Hours £	Materials £	O & P £	Total £
38mm	nr	0.03	0.31	0.20	0.08	0.59
50mm	nr	0.03	0.31	0.22	0.08	0.61
63mm	nr	0.03	0.31	0.24	0.08	0.63
75mm	nr	0.03	0.31	0.26	0.09	0.66

	Unit	Hours	Hours £	Materials £	O & P £	Total £
Galvanised mild steel square toothed timber connectors, double sided, diameter						
38mm	nr	0.03	0.31	0.22	0.08	0.61
50mm	nr	0.03	0.31	0.24	0.08	0.63
63mm	nr	0.03	0.31	0.26	0.09	0.66
75mm	nr	0.03	0.31	0.28	0.09	0.68
Galvanised mild steel restraint straps 30 x 2.5mm, girth						
400mm	nr	0.16	1.66	2.86	0.68	5.20
500mm	nr	0.18	1.87	3.10	0.75	5.72
600mm	nr	0.20	2.08	3.55	0.84	6.47
700mm	nr	0.22	2.29	3.87	0.92	7.08
800mm	nr	0.24	2.50	4.15	1.00	7.64
900mm	nr	0.28	2.91	4.66	1.14	8.71
1000mm	nr	0.30	3.12	4.89	1.20	9.21
Galvanised mild steel bolts, 100mm long, diameter						
6mm	nr	0.05	0.52	0.22	0.11	0.85
8mm	nr	0.05	0.52	0.28	0.12	0.92
10mm	nr	0.06	0.62	0.38	0.15	1.15
12mm	nr	0.06	0.62	0.52	0.17	1.32
16mm	nr	0.08	0.83	0.66	0.22	1.72
20mm	nr	0.08	0.83	0.87	0.26	1.96
Galvanised mild steel bolts, 140mm long, diameter						
6mm	nr	0.05	0.52	0.36	0.13	1.01
8mm	nr	0.05	0.52	0.42	0.14	1.08
10mm	nr	0.06	0.62	0.52	0.17	1.32
12mm	nr	0.06	0.62	0.66	0.19	1.48
16mm	nr	0.08	0.83	0.80	0.24	1.88
20mm	nr	0.08	0.83	1.00	0.27	2.11

	Unit	Hours	Hours £	Materials £	O & P £	Total £
Woodwool reinforced slabs 50mm thick, fixed to softwood joists in lengths						
1800mm	m2	0.70	7.28	15.65	3.44	26.37
2100mm	m2	0.70	7.28	15.65	3.44	26.37
2400mm	m2	0.70	7.28	16.32	3.54	27.14
2700mm	m2	0.70	7.28	16.48	3.56	27.32
3000mm	m2	0.70	7.28	16.48	3.56	27.32
Woodwool reinforced slabs 75mm thick, fixed to softwood joists in lengths						
1800mm	m2	0.75	7.80	22.04	4.48	34.32
2100mm	m2	0.75	7.80	22.04	4.48	34.32
2400mm	m2	0.75	7.80	22.90	4.61	35.31
2700mm	m2	0.75	7.80	23.24	4.66	35.70
3000mm	m2	0.75	7.80	23.24	4.66	35.70

Flooring

	Unit	Hours	Hours £	Materials £	O & P £	Total £
Wrought softwood boarding, butt-jointed						
19 x 125mm	m2	0.60	6.24	6.55	1.92	14.71
25 x 100mm	m2	0.65	6.76	8.24	2.25	17.25
25 x 150mm	m2	0.60	6.24	8.90	2.27	17.41
Wrought softwood boarding, tongued and grooved						
19 x 125mm	m2	0.70	7.28	7.05	2.15	16.48
25 x 100mm	m2	0.75	7.80	8.74	2.48	19.02
25 x 150mm	m2	0.70	7.28	9.40	2.50	19.18

	Unit	Hours	Hours £	Materials £	O & P £	Total £
Chipboard flooring, butt-jointed, thickness						
18mm	m2	0.35	3.64	3.48	1.07	8.19
25mm	m2	0.40	4.16	4.20	1.25	9.61
Chipboard flooring, tongued and grooved, thickness						
25mm	m2	0.50	5.20	4.46	1.45	11.11

Linings

	Unit	Hours	Hours £	Materials £	O & P £	Total £
Melamine-faced chipboard, 15mm thick to walls						
over 300mm wide	m2	1.10	11.44	4.27	2.36	18.07
less than 300mm wide	m	0.70	7.28	1.61	1.33	10.22
Plain chipboard, 12mm thick to walls						
over 300mm wide	m2	0.48	4.99	2.47	1.12	8.58
less than 300mm wide	m	0.28	2.91	0.78	0.55	4.25
Birch-faced blockboard, 12mm thick to walls						
over 300mm wide	m2	0.60	6.24	9.30	2.33	17.87
less than 300mm wide	m	0.40	4.16	3.10	1.09	8.35
Standard quality hardboard, 3.2mm thick to walls						
over 300mm wide	m2	0.40	4.16	1.90	0.91	6.97
less than 300mm wide	m	0.25	2.60	0.65	0.49	3.74

	Unit	Hours	Hours £	Materials £	O & P £	Total £
Standard quality hardboard, 6mm thick to walls						
over 300mm wide	m2	0.45	4.68	2.97	1.15	8.80
less than 300mm wide	m	0.30	3.12	1.02	0.62	4.76
Internal quality plywood, 4mm thick to walls						
over 300mm wide	m2	0.50	5.20	2.78	1.20	9.18
less than 300mm wide	m	0.30	3.12	1.15	0.64	4.91
Internal quality plywood, 6mm thick to walls						
over 300mm wide	m2	0.55	5.72	3.12	1.33	10.17
less than 300mm wide	m	0.34	3.54	1.28	0.72	5.54
Insulation board, 12mm thick to walls						
over 300mm wide	m2	0.40	4.16	1.92	0.91	6.99
less than 300mm wide	m	0.24	2.50	0.73	0.48	3.71

Windows

Standard softwood windows without glazing bars, type

	Unit	Hours	Hours £	Materials £	O & P £	Total £
N07V, 488 x 750mm	nr	0.75	7.80	42.23	7.50	57.53
N09V, 488 x 900mm	nr	1.00	10.40	44.32	8.21	62.93
N12V, 488 x 1200mm	nr	1.25	13.00	48.51	9.23	70.74
107C, 630 x 750mm	nr	0.75	7.80	51.37	8.88	68.05
110C, 630 x 1050mm	nr	1.00	10.40	53.69	9.61	73.70
112C, 630 x 1200mm	nr	1.25	13.00	55.50	10.28	78.78
109V, 630 x 900mm	nr	0.75	7.80	54.36	9.32	71.48
110V, 630 x 1050mm	nr	1.25	13.00	56.44	10.42	79.86
112V, 630 x 1200mm	nr	0.75	7.80	57.88	9.85	75.53
2N09W, 915 x 900mm	nr	0.90	9.36	58.97	10.25	78.58

	Unit	Hours	Hours £	Materials £	O & P £	Total £
2N10W, 915 x 1050mm	nr	1.15	11.96	61.23	10.98	84.17
2N12W, 915 x 1200mm	nr	1.40	14.56	63.48	11.71	89.75
212W, 1200 x 1200mm	nr	1.40	14.56	75.66	13.53	103.75
210C, 1200 x 1050mm	nr	1.40	14.56	68.97	12.53	96.06
212T, 1200 x 1200mm	nr	1.40	14.56	84.37	14.84	113.77

Standard hardwood windows
without glazing bars, type

	Unit	Hours	Hours £	Materials £	O & P £	Total £
H2N10W, 915 x 1050mm	nr	1.00	10.40	116.37	19.02	145.79
H2N13W, 915 x 1050mm	nr	1.25	13.00	122.54	20.33	155.87
H2N15W, 915 x 1500mm	nr	1.50	15.60	127.68	21.49	164.77
H213W, 1200 x 1350mm	nr	1.60	16.64	141.27	23.69	181.60
H215W, 1200 x 1500mm	nr	1.70	17.68	145.69	24.51	187.88
H310CC, 1770 x 1050mm	nr	2.00	20.80	214.55	35.30	270.65
H312CC, 1770 x 1200mm	nr	2.20	22.88	232.44	38.30	293.62

PVC-U windows, single glazed,
size

	Unit	Hours	Hours £	Materials £	O & P £	Total £
600 x 900mm	nr	1.00	10.40	48.56	8.84	67.80
600 x 1200mm	nr	1.30	13.52	92.47	15.90	121.89
1200 x 1200mm	nr	1.60	16.64	115.41	19.81	151.86
1800 x 1200mm	nr	2.00	20.80	131.64	22.87	175.31

PVC-U windows, double glazed,
size

	Unit	Hours	Hours £	Materials £	O & P £	Total £
600 x 900mm	nr	1.00	10.40	62.87	10.99	84.26
600 x 1200mm	nr	1.30	13.52	108.74	18.34	140.60
1200 x 1200mm	nr	1.60	16.64	149.96	24.99	191.59
1800 x 1200mm	nr	2.00	20.80	175.32	29.42	225.54

	Unit	Hours	Hours £	Materials £	O & P £	Total £

Doors

Standard internal flush doors,
hardboard faced both sides,
35mm thick, size

	Unit	Hours	Hours £	Materials £	O & P £	Total £
533 x 1981mm	nr	1.20	12.48	23.25	5.36	41.09
610 x 1981mm	nr	1.20	12.48	23.25	5.36	41.09
686 x 1981mm	nr	1.20	12.48	23.25	5.36	41.09
762 x 1981mm	nr	1.20	12.48	23.25	5.36	41.09

Standard internal flush doors,
hardboard faced both sides,
40mm thick, size

	Unit	Hours	Hours £	Materials £	O & P £	Total £
533 x 1981mm	nr	1.20	12.48	25.12	5.64	43.24
610 x 1981mm	nr	1.20	12.48	25.12	5.64	43.24
686 x 1981mm	nr	1.20	12.48	25.12	5.64	43.24
762 x 1981mm	nr	1.20	12.48	25.12	5.64	43.24

Standard internal flush doors,
sapele faced both sides, 35mm
thick, size

	Unit	Hours	Hours £	Materials £	O & P £	Total £
533 x 1981mm	nr	1.30	13.52	35.14	7.30	55.96
610 x 1981mm	nr	1.30	13.52	35.14	7.30	55.96
686 x 1981mm	nr	1.30	13.52	35.14	7.30	55.96
762 x 1981mm	nr	1.30	13.52	35.14	7.30	55.96

Standard internal flush doors,
sapele faced both sides, 40mm
thick, size

	Unit	Hours	Hours £	Materials £	O & P £	Total £
533 x 1981mm	nr	1.30	13.52	38.88	7.86	60.26
610 x 1981mm	nr	1.30	13.52	38.88	7.86	60.26
686 x 1981mm	nr	1.30	13.52	38.88	7.86	60.26
762 x 1981mm	nr	1.30	13.52	38.88	7.86	60.26

	Unit	Hours	Hours £	Materials £	O & P £	Total £
Standard internal flush half-hour fire check doors, hardboard faced both sides, 44mm thick, size						
686 x 1981mm	nr	1.60	16.64	54.23	10.63	81.50
762 x 1981mm	nr	1.60	16.64	55.32	10.79	82.75
726 x 2040mm	nr	1.60	16.64	57.88	11.18	85.70
826 x 2040mm	nr	1.60	16.64	58.39	11.25	86.28
Standard internal flush half-hour fire check doors, sapele faced both sides, 44mm thick, size						
686 x 1981mm	nr	1.80	18.72	74.55	13.99	107.26
762 x 1981mm	nr	1.80	18.72	76.32	14.26	109.30
726 x 2040mm	nr	1.80	18.72	78.97	14.65	112.34
826 x 2040mm	nr	1.80	18.72	80.02	14.81	113.55
Standard external flush doors, plywood faced both sides, 54mm thick, size						
762 x 1981mm	nr	2.10	21.84	52.56	11.16	85.56
838 x 1981mm	nr	2.10	21.84	54.74	11.49	88.07
Framed, ledged and braced doors, 44mm thick						
726 x 2040mm	nr	1.30	13.52	68.85	12.36	94.73
826 x 2040mm	nr	1.30	13.52	70.52	12.61	96.65

Door frames and linings

	Unit	Hours	Hours £	Materials £	O & P £	Total £
Standard door lining with loose stops, for door size						
686 x 1981mm						
27 x 94mm	nr	0.75	7.80	32.63	6.06	46.49
27 x 107mm	nr	0.75	7.80	34.21	6.30	48.31

	Unit	Hours	Hours £	Materials £	O & P £	Total £
27 x 121mm	nr	0.75	7.80	35.89	6.55	50.24
27 x 133mm	nr	0.75	7.80	37.87	6.85	52.52
762 x 1981mm						
27 x 94mm	nr	0.75	7.80	33.25	6.16	47.21
27 x 107mm	nr	0.75	7.80	35.47	6.49	49.76
27 x 121mm	nr	0.75	7.80	37.10	6.74	51.64
27 x 133mm	nr	0.75	7.80	38.67	6.97	53.44
762 x 2040mm						
27 x 94mm	nr	0.75	7.80	37.43	6.78	52.01
27 x 107mm	nr	0.75	7.80	39.41	7.08	54.29
27 x 121mm	nr	0.75	7.80	42.64	7.57	58.01
27 x 133mm	nr	0.75	7.80	43.91	7.76	59.47

Standard rebated door frame, for door size

	Unit	Hours	Hours £	Materials £	O & P £	Total £
686 x 1981mm						
35 x 107mm	nr	0.75	7.80	34.21	6.30	48.31
35 x 133mm	nr	0.75	7.80	34.21	6.30	48.31
762 x 1981mm						
35 x 107mm	nr	0.75	7.80	34.21	6.30	48.31
35 x 133mm	nr	0.75	7.80	34.21	6.30	48.31

Wrought softwood rebated and rounded frames, size

	Unit	Hours	Hours £	Materials £	O & P £	Total £
38 x 100mm	m	0.24	2.50	5.78	1.24	9.52
38 x 125mm	m	0.24	2.50	7.51	1.50	11.51
38 x 140mm	m	0.24	2.50	8.79	1.69	12.98
63 x 88mm	m	0.26	2.70	9.22	1.79	13.71
63 x 100mm	m	0.26	2.70	9.56	1.84	14.10
63 x 125mm	m	0.26	2.70	10.76	2.02	15.48
75 x 100mm	m	0.28	2.91	9.67	1.89	14.47
75 x 125mm	m	0.28	2.91	14.00	2.54	19.45
75 x 140mm	m	0.28	2.91	15.27	2.73	20.91

	Unit	Hours	Hours £	Materials £	O & P £	Total £
Mahogany rebated and rounded frames, size						
38 x 100mm	m	0.30	3.12	12.44	2.33	17.89
38 x 125mm	m	0.30	3.12	14.20	2.60	19.92
38 x 140mm	m	0.30	3.12	16.27	2.91	22.30
63 x 88mm	m	0.32	3.33	17.84	3.18	24.34
63 x 100mm	m	0.32	3.33	18.57	3.28	25.18
63 x 125mm	m	0.32	3.33	20.78	3.62	27.72
75 x 100mm	m	0.36	3.74	20.44	3.63	27.81
75 x 125mm	m	0.36	3.74	27.51	4.69	35.94
75 x 140mm	m	0.36	3.74	33.45	5.58	42.77

Stairs

Wrought softwood open-tread staircase in one flight, 2600mm height, 2700mm going, 12 treads and 13 risers, width

	Unit	Hours	Hours £	Materials £	O & P £	Total £
850 mm	nr	14.00	145.60	342.45	73.21	561.26
910 mm	nr	14.00	145.60	362.77	76.26	584.63

Wrought softwood closed-tread staircase in one flight, 2600mm height, 2700mm going, 12 treads and 13 risers, width

	Unit	Hours	Hours £	Materials £	O & P £	Total £
850 mm	nr	16.00	166.40	486.37	97.92	750.69
910 mm	nr	16.00	166.40	514.81	102.18	783.39

Hardwood open-tread staircase in one flight, 2600mm height, 2700mm going, 12 treads and 13 risers, width

	Unit	Hours	Hours £	Materials £	O & P £	Total £
850 mm	nr	14.00	145.60	634.55	117.02	897.17
910 mm	nr	14.00	145.60	678.51	123.62	947.73

	Unit	Hours	Hours £	Materials £	O & P £	Total £
Hardwood closed-tread staircase in one flight, 2600mm height, 2700mm going, 12 treads and 13 risers, width						
850 mm	nr	16.00	166.40	789.40	143.37	1099.17
910 mm	nr	16.00	166.40	828.27	149.20	1143.87

Kitchen fittings

Fix only kitchen fittings

wall units 200mm high, size						
300 x 600mm	nr	1.00	10.40	0.00	1.56	11.96
300 x 1000mm	nr	1.10	11.44	0.00	1.72	13.16
300 x 1200mm	nr	1.20	12.48	0.00	1.87	14.35
base units 750mm high, size						
600 x 900mm	nr	1.20	12.48	0.00	1.87	14.35
600 x 1000mm	nr	1.25	13.00	0.00	1.95	14.95
600 x 1200mm	nr	1.30	13.52	0.00	2.03	15.55
900 x 900mm	nr	1.30	13.52	0.00	2.03	15.55
900 x 1000mm	nr	1.35	14.04	0.00	2.11	16.15
900 x 1200mm	nr	1.40	14.56	0.00	2.18	16.74
sink units 750mm high, size						
600 x 900mm	nr	1.30	13.52	0.00	2.03	15.55
600 x 1000mm	nr	1.35	14.04	0.00	2.11	16.15
600 x 1200mm	nr	1.40	14.56	0.00	2.18	16.74
worktops, size						
600 x 900mm	nr	0.40	4.16	0.00	0.62	4.78
600 x 1000mm	nr	0.50	5.20	0.00	0.78	5.98
600 x 1200mm	nr	0.60	6.24	0.00	0.94	7.18

	Unit	Hours	Hours £	Materials £	O & P £	Total £

Insulation

Glass fibre quilt laid over ceiling joists, thickness

	Unit	Hours	Hours £	Materials £	O & P £	Total £
50mm	m2	0.15	1.56	2.10	0.55	4.21
80mm	m2	0.17	1.77	2.41	0.63	4.80
100mm	m2	0.20	2.08	3.36	0.82	6.26
150mm	m2	0.24	2.50	5.27	1.16	8.93

Glass fibre quilt fixed vertically to softwood, thickness

	Unit	Hours	Hours £	Materials £	O & P £	Total £
50mm	m2	0.20	2.08	2.10	0.63	4.81
80mm	m2	0.23	2.39	2.41	0.72	5.52

Expanded polystyrene board fixed vertically to walls with adhesive, thickness

	Unit	Hours	Hours £	Materials £	O & P £	Total £
12mm	m2	0.50	5.20	1.55	1.01	7.76
25mm	m2	0.60	6.24	2.87	1.37	10.48
50mm	m2	0.70	7.28	4.65	1.79	13.72

Ironmongery

Fix only to softwood

door furniture

	Unit	Hours	Hours £	Materials £	O & P £	Total £
light butts	nr	0.30	3.12	0.00	0.47	3.59
medium butts	nr	0.33	3.43	0.00	0.51	3.95
heavy butts	nr	0.35	3.64	0.00	0.55	4.19
rising butts	nr	0.40	4.16	0.00	0.62	4.78
tee bands	nr	1.00	10.40	0.00	1.56	11.96
barrel bolts, small	nr	0.55	5.72	0.00	0.86	6.58
barrel bolts, medium	nr	0.60	6.24	0.00	0.94	7.18
barrel bolts, large	nr	0.70	7.28	0.00	1.09	8.37
tower bolts, small	nr	0.55	5.72	0.00	0.86	6.58
tower bolts, medium	nr	0.60	6.24	0.00	0.94	7.18
tower bolts, large	nr	0.70	7.28	0.00	1.09	8.37

	Unit	Hours	Hours £	Materials £	O & P £	Total £
panic bolts	nr	2.20	22.88	0.00	3.43	26.31
overhead door closer, medium	nr	1.00	10.40	0.00	1.56	11.96
overhead door closer, heavy	nr	1.25	13.00	0.00	1.95	14.95
mortice latch	nr	0.80	8.32	0.00	1.25	9.57
cylinder night rim latch	nr	1.00	10.40	0.00	1.56	11.96
rim dead lock	nr	0.70	7.28	0.00	1.09	8.37
mortice dead lock	nr	0.90	9.36	0.00	1.40	10.76
mortice latch furniture	nr	0.30	3.12	0.00	0.47	3.59
Suffolk latch	nr	0.65	6.76	0.00	1.01	7.77
Norfolk latch	nr	0.65	6.76	0.00	1.01	7.77
postal knocker and plate	nr	1.00	10.40	0.00	1.56	11.96
pull handles	nr	0.25	2.60	0.00	0.39	2.99
push plates	nr	0.25	2.60	0.00	0.39	2.99
window furniture						
casement stay	nr	0.35	3.64	0.00	0.55	4.19
casement fastener	nr	0.25	2.60	0.00	0.39	2.99
cockspur fastener	nr	0.30	3.12	0.00	0.47	3.59
sundries						
numerals	nr	0.05	0.52	0.00	0.08	0.60
hat and coat hook	nr	0.05	0.52	0.00	0.08	0.60
toilet roll holder	nr	0.10	1.04	0.00	0.16	1.20
cabin hook	nr	0.05	0.52	0.00	0.08	0.60
Fix only to hardwood						
door furniture						
light butts	nr	0.40	4.16	0.00	0.62	4.78
medium butts	nr	0.43	4.47	0.00	0.67	5.14
heavy butts	nr	0.45	4.68	0.00	0.70	5.38
rising butts	nr	0.50	5.20	0.00	0.78	5.98
tee bands	nr	1.15	11.96	0.00	1.79	13.75
barrel bolts, small	nr	0.75	7.80	0.00	1.17	8.97
barrel bolts, medium	nr	0.70	7.28	0.00	1.09	8.37
barrel bolts, large	nr	0.80	8.32	0.00	1.25	9.57
tower bolts, small	nr	0.65	6.76	0.00	1.01	7.77
tower bolts, medium	nr	0.70	7.28	0.00	1.09	8.37
tower bolts, large	nr	0.80	8.32	0.00	1.25	9.57
panic bolts	nr	2.40	24.96	0.00	3.74	28.70

	Unit	Hours	Hours £	Materials £	O & P £	Total £
overhead door closer, medium	nr	1.15	11.96	0.00	1.79	13.75
overhead door closer, heavy	nr	1.40	14.56	0.00	2.18	16.74
mortice latch	nr	0.90	9.36	0.00	1.40	10.76
cylinder night rim latch	nr	1.15	11.96	0.00	1.79	13.75
rim dead lock	nr	0.80	8.32	0.00	1.25	9.57
mortice dead lock	nr	1.00	10.40	0.00	1.56	11.96
mortice latch furniture	nr	0.40	4.16	0.00	0.62	4.78
Suffolk latch	nr	0.75	7.80	0.00	1.17	8.97
Norfolk latch	nr	0.75	7.80	0.00	1.17	8.97
postal knocker and plate	nr	1.15	11.96	0.00	1.79	13.75
pull handles	nr	0.30	3.12	0.00	0.47	3.59
push plates	nr	0.30	3.12	0.00	0.47	3.59
window furniture						
casement stay	nr	0.45	4.68	0.00	0.70	5.38
casement fastener	nr	0.35	3.64	0.00	0.55	4.19
cockspur fastener	nr	0.40	4.16	0.00	0.62	4.78
sundries						
numerals	nr	0.07	0.73	0.00	0.11	0.84
hat and coat hook	nr	0.07	0.73	0.00	0.11	0.84
toilet roll holder	nr	0.14	1.46	0.00	0.22	1.67
cabin hook	nr	0.07	0.73	0.00	0.11	0.84

	Unit	Hours	Hours £	Materials £	O & P £	Total £

METALWORK

Balustrades

Galvanised steel balustrades

50 x 8mm flat bar	m	0.50	5.06	18.37	3.51	26.94
welded angles	nr	0.75	7.58	0.00	1.14	8.72
ramps	nr	0.75	7.58	0.00	1.14	8.72
flat bends	nr	0.75	7.58	0.00	1.14	8.72
38 x 12mm rails	m	0.50	5.06	18.37	3.51	26.94
welded angles	nr	0.75	7.58	0.00	1.14	8.72
ramps	nr	0.75	7.58	0.00	1.14	8.72
flat bends	nr	0.75	7.58	0.00	1.14	8.72

Lintels

Standard galvanised steel lintels to
100mm thick wall, 143mm high

900mm long	nr	0.15	1.52	5.57	1.06	8.15
1050mm long	nr	0.20	2.02	5.97	1.20	9.19
1200mm long	nr	0.25	2.53	6.55	1.36	10.44

Standard galvanised steel lintels to
256mm thick wall, 143mm high

900mm long	nr	0.20	2.02	31.24	4.99	38.25
1500mm long	nr	0.25	2.53	48.36	7.63	58.52
2100mm long	nr	0.30	3.03	69.52	10.88	83.44

Standard galvanised steel lintels to
256mm thick wall, 219mm high

2700mm long	nr	0.75	7.58	98.46	15.91	121.95
4200mm long	nr	1.00	10.11	233.57	36.55	280.23

	Unit	Hours	Hours £	Materials £	O & P £	Total £

Sundries

Galvanised mild steel water bars, size

	Unit	Hours	Hours £	Materials £	O & P £	Total £
5 x 25mm	m	0.50	5.06	2.68	1.16	8.90
5 x 30mm	m	0.50	5.06	3.22	1.24	9.52
5 x 40mm	m	0.50	5.06	4.05	1.37	10.47

Galvanised steel dowels, size

	Unit	Hours	Hours £	Materials £	O & P £	Total £
8mm diameter x 50mm	nr	0.10	1.01	0.20	0.18	1.39
8mm diameter x 100mm	nr	0.12	1.21	0.26	0.22	1.69
10mm diameter x 50mm	nr	0.10	1.01	0.24	0.19	1.44
10mm diameter x 100mm	nr	0.12	1.21	0.30	0.23	1.74
12mm diameter x 50mm	nr	0.10	1.01	0.26	0.19	1.46
12mm diameter x 100mm	nr	0.12	1.21	0.32	0.23	1.76

	Unit	Hours	Hours £	Materials £	O & P £	Total £

ROOFING

Tiling

Marley Plain granuled or smooth
finish tiles size 267 x 165mm,
65mm lap, 35 degree pitch, type
1F reinforced underlay

battens size 38 x 19mm						
gauge 100mm	m2	1.35	13.65	23.48	5.57	42.70
gauge 95mm	m2	1.40	14.15	24.56	5.81	44.52
gauge 90mm	m2	1.45	14.66	25.87	6.08	46.61
battens size 38 x 25mm						
gauge 100mm	m2	1.35	13.65	23.78	5.61	43.04
gauge 95mm	m2	1.40	14.15	24.86	5.85	44.87
gauge 90mm	m2	1.45	14.66	26.17	6.12	46.95

Extra for

nailing every tile with aluminium nails	m2	0.15	1.52	0.40	0.29	2.20
interlocking dry verge system	m	0.20	2.02	9.78	1.77	13.57
verge, 150mm wide plain tile undercloak	m	0.20	2.02	1.95	0.60	4.57
double course at eaves	m	0.35	3.54	3.22	1.01	7.77
segmental ridge tile	m	0.40	4.04	8.87	1.94	14.85
valley trough tiles	m	0.60	6.07	13.28	2.90	22.25
segmental hip tiles	m	0.60	6.07	13.28	2.90	22.25
bonnet hip tiles	m	0.60	6.07	13.28	2.90	22.25
eaves vent system	m	0.50	5.06	14.32	2.91	22.28
ventilated ridge terminal	nr	0.50	5.06	33.62	5.80	44.48
gas vent terminal	nr	0.50	5.06	49.57	8.19	62.82
soil vent terminal	nr	0.50	5.06	31.49	5.48	42.03
straight cutting	m	0.20	2.02	0.00	0.30	2.33
holes for pipes	nr	0.40	4.04	0.00	0.61	4.65

	Unit	Hours	Hours £	Materials £	O & P £	Total £
Marley Ludlow Plus smooth finish tiles size 387 x 229mm, battens size 38 x 25mm, type 1F reinforced underlay						
75mm lap, pitch 25-44 degrees	m2	0.90	9.10	10.65	2.96	22.71
100mm lap, pitch 22-44 degrees	m2	0.95	9.60	10.76	3.05	23.42
Extra for						
nailing every tile with aluminium nails	m2	0.10	1.01	0.30	0.20	1.51
interlocking dry verge system	m	0.20	2.02	9.78	1.77	13.57
verge, 150mm wide plain tile undercloak	m	0.20	2.02	1.95	0.60	4.57
segmental ridge tile	m	0.40	4.04	8.87	1.94	14.85
valley trough tiles	m	0.60	6.07	13.28	2.90	22.25
segmental hip tiles	m	0.60	6.07	13.28	2.90	22.25
eaves vent system	m	0.50	5.06	14.32	2.91	22.28
ventilated ridge terminal	nr	0.50	5.06	33.62	5.80	44.48
gas vent terminal	nr	0.50	5.06	49.57	8.19	62.82
soil vent terminal	nr	0.50	5.06	31.49	5.48	42.03
straight cutting	m	0.20	2.02	0.00	0.30	2.33
holes for pipes	nr	0.40	4.04	0.00	0.61	4.65
Marley Modern smooth finish tiles size 420 x 330mm, battens size 38 x 25mm, type 1F reinforced underlay						
75mm lap, pitch 22.5-44 degrees	m2	0.80	8.09	10.98	2.86	21.93
Extra for						
nailing every tile with aluminium nails	m2	0.05	0.51	0.20	0.11	0.81
verge, 150mm wide plain tile undercloak	m	0.20	2.02	1.95	0.60	4.57
interlocking dry verge system	m	0.20	2.02	9.78	1.77	13.57
Modern ridge tile	m	0.60	6.07	9.57	2.35	17.98

	Unit	Hours	Hours £	Materials £	O & P £	Total £
Modern monoridge	m	0.60	6.07	12.88	2.84	21.79
dry ridge system	m	0.50	5.06	11.56	2.49	19.11
Modern hip tiles	m	0.60	6.07	8.99	2.26	17.31
eaves vent system	m	0.50	5.06	14.32	2.91	22.28
ventilated ridge terminal	nr	0.50	5.06	33.62	5.80	44.48
gas vent terminal	nr	0.50	5.06	49.57	8.19	62.82
soil vent terminal	nr	0.50	5.06	31.49	5.48	42.03
straight cutting	m	0.20	2.02	0.00	0.30	2.33
holes for pipes	nr	0.40	4.04	0.00	0.61	4.65

Redland Renown granular faced
or through-coloured tiles size 418
x 330mm, 75mm lap, 343mm
gauge, pitch 30-40 degrees, type
1F reinforced underlay

	Unit	Hours	Hours £	Materials £	O & P £	Total £
battens size 38 x 22mm	m2	0.80	8.09	10.58	2.80	21.47
battens size 38 x 25mm	m2	0.85	8.59	10.67	2.89	22.15

Extra for

	Unit	Hours	Hours £	Materials £	O & P £	Total £
nailing every tile with aluminium nails	m2	0.05	0.51	0.20	0.11	0.81
cloaked verge tile	m	0.20	2.02	8.12	1.52	11.66
half-round ridge or hip tile	m	0.60	6.07	11.75	2.67	20.49
gas flue ridge terminal	nr	0.80	8.09	93.68	15.27	117.03
valley trough tiles	m	0.60	6.07	16.37	3.37	25.80
straight cutting	m	0.20	2.02	0.00	0.30	2.33
holes for pipes	nr	0.40	4.04	0.00	0.61	4.65

Redland Norfolk granular
through-coloured tiles size 381
x 227mm, 100mm lap, 306mm
gauge, pitch 22.5-25 degrees, type
1F reinforced underlay

	Unit	Hours	Hours £	Materials £	O & P £	Total £
battens size 38 x 22mm	m2	0.85	8.59	13.87	3.37	25.83
battens size 38 x 25mm	m2	0.90	9.10	14.02	3.47	26.59

	Unit	Hours	Hours £	Materials £	O & P £	Total £
Extra for						
nailing every tile with aluminium nails	m2	0.05	0.51	0.20	0.11	0.81
dryvent ridge	m	0.75	7.58	14.88	3.37	25.83
half-round ridge or hip tile	m	0.60	6.07	11.75	2.67	20.49
gas flue ridge terminal	nr	0.80	8.09	91.45	14.93	114.47
valley trough tiles	m	0.60	6.07	16.78	3.43	26.27
straight cutting	m	0.20	2.02	0.00	0.30	2.33
holes for pipes	nr	0.40	4.04	0.00	0.61	4.65

Fibre-cement slating

Asbestos-free artificial slates size
400 x 200mm, pitch 40-45 degrees,
38 x 25mm softwood battens, type
1F reinforced underlay

	Unit	Hours	Hours £	Materials £	O & P £	Total £
lap 70mm, gauge 165mm	m2	1.10	11.12	29.77	6.13	47.02

Asbestos-free artificial slates size
400 x 200mm, pitch 40-45 degrees,
38 x 25mm softwood battens, type
1F reinforced underlay

	Unit	Hours	Hours £	Materials £	O & P £	Total £
lap 90mm, gauge 155mm	m2	1.15	11.63	30.24	6.28	48.15

Asbestos-free artificial slates size
500 x 250mm, pitch over 25 degrees,
38 x 25mm softwood battens, type
1F reinforced underlay

	Unit	Hours	Hours £	Materials £	O & P £	Total £
lap 90mm, gauge 205mm	m2	1.00	10.11	27.51	5.64	43.26

Asbestos-free artificial slates size
600 x 300mm, pitch over 25 degrees,
38 x 25mm softwood battens, type
1F reinforced underlay

	Unit	Hours	Hours £	Materials £	O & P £	Total £
lap 100mm, gauge 250mm	m2	0.90	9.10	24.01	4.97	38.08

	Unit	Hours	Hours £	Materials £	O & P £	Total £
Extra for						
verge, 150mm wide plain tile						
undercloak	m	0.20	2.02	1.88	0.59	4.49
double course at eaves	m	0.35	3.54	5.90	1.42	10.85
half-round ridge tile	m	0.40	4.04	21.24	3.79	29.08
valley tiles	m	0.60	6.07	21.24	4.10	31.40
straight cutting	m	0.20	2.02	0.00	0.30	2.33
holes for pipes	nr	0.40	4.04	0.00	0.61	4.65

Natural slating

Blue/grey Welsh slates size 405 x
205mm, 75mm lap, 50 x 25mm
softwood battens, type 1F
reinforced underlay

	Unit	Hours	Hours £	Materials £	O & P £	Total £
sloping	m2	1.20	12.13	47.22	8.90	68.25
vertical	m2	1.40	14.15	47.22	9.21	70.58

Extra for

	Unit	Hours	Hours £	Materials £	O & P £	Total £
double eaves course	m	0.50	5.06	22.13	4.08	31.26
single verge undercloak course	m	0.70	7.08	16.57	3.55	27.19
angled ridge or hip tiles	m	0.70	7.08	14.37	3.22	24.66
mitred hips including cutting						
both sides	m	0.70	7.08	24.36	4.72	36.15
straight cutting	m	0.60	6.07	0.00	0.91	6.98
holes for pipes	nr	0.40	4.04	0.00	0.61	4.65
fix only lead soakers	nr	0.50	5.06	0.00	0.76	5.81

Blue/grey Welsh slates size 510 x
255mm, 75mm lap, 50 x 25mm
softwood battens, type 1F
reinforced underlay

	Unit	Hours	Hours £	Materials £	O & P £	Total £
sloping	m2	1.00	10.11	49.21	8.90	68.22
vertical	m2	1.20	12.13	49.21	9.20	70.54

	Unit	Hours	Hours £	Materials £	O & P £	Total £
Extra for						
double eaves course	m	0.50	5.06	27.68	4.91	37.65
single verge undercloak course	m	0.70	7.08	17.58	3.70	28.36
angled ridge or hip tiles	m	0.70	7.08	14.37	3.22	24.66
mitred hips including cutting						
both sides	m	0.70	7.08	24.36	4.72	36.15
straight cutting	m	0.60	6.07	0.00	0.91	6.98
holes for pipes	nr	0.40	4.04	0.00	0.61	4.65
fix only lead soakers	nr	0.50	5.06	0.00	0.76	5.81
Blue/grey Welsh slates size 610 x 3005mm, 75mm lap, 50 x 25mm softwood battens, type 1F reinforced underlay						
sloping	m2	0.80	8.09	52.39	9.07	69.55
vertical	m2	1.00	10.11	52.39	9.38	71.88
Extra for						
double eaves course	m	0.50	5.06	31.24	5.44	41.74
single verge undercloak course	m	0.70	7.08	19.27	3.95	30.30
angled ridge or hip tiles	m	0.70	7.08	14.37	3.22	24.66
mitred hips including cutting						
both sides	m	0.70	7.08	24.36	4.72	36.15
straight cutting	m	0.60	6.07	0.00	0.91	6.98
holes for pipes	nr	0.40	4.04	0.00	0.61	4.65
fix only lead soakers	nr	0.50	5.06	0.00	0.76	5.81
Westmorland green slates in random lengths 450 - 230mm, 75mm laps, 50 x 25mm softwood battens, type 1F reinforced underlay						
sloping	m2	1.35	13.65	105.55	17.88	137.08
vertical	m2	1.55	15.67	105.55	18.18	139.40

	Unit	Hours	Hours £	Materials £	O & P £	Total £
Extra for						
double eaves course	m	0.50	5.06	31.24	5.44	41.74
single verge undercloak course	m	0.70	7.08	19.27	3.95	30.30
angled ridge or hip tiles	m	0.70	7.08	14.37	3.22	24.66
straight cutting	m	0.60	6.07	0.00	0.91	6.98
holes for pipes	nr	0.40	4.04	0.00	0.61	4.65
fix only lead soakers	nr	0.50	5.06	0.00	0.76	5.81

Reconstructed stone slating

Marley Monarch interlocking slate
size 325 x 330mm, 38 x 25mm
softwood battens, type 1F
reinforced underlay

	Unit	Hours	Hours £	Materials £	O & P £	Total £
75mm lap, pitch 25-90 degrees	m2	1.00	10.11	28.46	5.79	44.36
100mm lap, pitch 25-90 degrees	m2	1.05	10.62	28.99	5.94	45.55
nailing every tile with aluminium nails	m2	0.05	0.51	0.20	0.11	0.81
interlocking dry verge system	m	0.20	2.02	9.78	1.77	13.57
verge, 150mm wide plain tile undercloak	m	0.20	2.02	1.95	0.60	4.57
Modern ridge tiles	m	0.60	6.07	8.99	2.26	17.31
Modern hip tiles	m	0.60	6.07	8.99	2.26	17.31
segmental ridge tile	m	0.40	4.04	8.87	1.94	14.85
Modern monoridge	m	0.60	6.07	12.88	2.84	21.79
dry ridge system	m	0.50	5.06	11.56	2.49	19.11
valley trough tiles	m	0.60	6.07	13.28	2.90	22.25
segmental hip tiles	m	0.60	6.07	13.28	2.90	22.25
ventilated ridge terminal	nr	0.50	5.06	33.62	5.80	44.48
gas vent terminal	nr	0.50	5.06	49.57	8.19	62.82
soil vent terminal	nr	0.50	5.06	31.49	5.48	42.03
straight cutting	m	0.20	2.02	0.00	0.30	2.33
holes for pipes	nr	0.40	4.04	0.00	0.61	4.65

	Unit	Hours	Hours £	Materials £	O & P £	Total £
Redland Cambrian through coloured slates size 300 x 336mm, 38 x 25mm softwood battens, type 1F reinforced underlay						
50mm lap, 250mm gauge	m2	0.80	8.09	26.34	5.16	39.59
90mm lap, 210mm gauge	m2	0.85	8.59	26.77	5.30	40.67
Extra for						
nailing every tile with aluminium nails	m2	0.05	0.51	0.20	0.11	0.81
Extra for						
nailing every tile with aluminium nails	m2	0.05	0.51	0.20	0.11	0.81
dryvent ridge	m	0.75	7.58	14.88	3.37	25.83
half-round ridge or hip tile	m	0.60	6.07	11.75	2.67	20.49
gas flue ridge terminal	nr	0.80	8.09	91.45	14.93	114.47
valley trough tiles	m	0.60	6.07	16.78	3.43	26.27
straight cutting	m	0.20	2.02	0.00	0.30	2.33
holes for pipes	nr	0.40	4.04	0.00	0.61	4.65
Lead sheet coverings						
Milled lead roof coverings to flat roofing						
code 5	m2	4.20	42.46	33.45	11.39	87.30
code 6	m2	4.40	44.48	35.68	12.02	92.19
Milled lead roof coverings to dormers						
code 5	m2	4.80	48.53	33.45	12.30	94.27
code 6	m2	5.00	50.55	35.68	12.93	99.16

	Unit	Hours	Hours £	Materials £	O & P £	Total £
Milled lead roof coverings sloping roofing						
code 5	m2	4.60	46.51	33.45	11.99	91.95
code 6	m2	4.80	48.53	35.68	12.63	96.84
Lead flashings, code 5						
horizontal						
150mm girth	m	0.45	4.55	5.02	1.44	11.00
200mm girth	m	0.55	5.56	6.68	1.84	14.08
300mm girth	m	0.60	6.07	10.04	2.42	18.52
sloping						
150mm girth	m	0.55	5.56	5.02	1.59	12.17
200mm girth	m	0.60	6.07	6.68	1.91	14.66
300mm girth	m	0.65	6.57	10.04	2.49	19.10
Lead aprons, code 5						
horizontal						
200mm girth	m	0.60	6.07	6.68	1.91	14.66
300mm girth	m	0.65	6.57	10.04	2.49	19.10
400mm girth	m	0.70	7.08	13.38	3.07	23.53
sloping						
200mm girth	m	0.65	6.57	6.68	1.99	15.24
300mm girth	m	0.70	7.08	10.04	2.57	19.68
400mm girth	m	0.75	7.58	13.38	3.14	24.11
Lead sills, code 5						
horizontal						
200mm girth	m	0.60	6.07	6.68	1.91	14.66
300mm girth	m	0.65	6.57	10.04	2.49	19.10
400mm girth	m	0.70	7.08	13.38	3.07	23.53

	Unit	Hours	Hours £	Materials £	O & P £	Total £
sloping						
200mm girth	m	0.65	6.57	6.68	1.99	15.24
300mm girth	m	0.70	7.08	10.04	2.57	19.68
400mm girth	m	0.75	7.58	13.38	3.14	24.11
Lead hips, code 5						
sloping						
200mm girth	m	0.70	7.08	6.68	2.06	15.82
300mm girth	m	0.75	7.58	10.04	2.64	20.27
400mm girth	m	0.80	8.09	13.38	3.22	24.69
Lead kerbs, code 5						
horizontal						
300mm girth	m	0.65	6.57	10.04	2.49	19.10
400mm girth	m	0.70	7.08	13.38	3.07	23.53
Lead valleys, code 5						
sloping						
400mm girth	m	0.80	8.09	13.38	3.22	24.69
600mm girth	m	0.85	8.59	20.07	4.30	32.96
800mm girth	m	0.90	9.10	26.76	5.38	41.24
Lead gutters, code 5						
sloping						
600mm girth	m	0.80	8.09	20.07	4.22	32.38
800mm girth	m	0.85	8.59	26.76	5.30	40.66
Slates, size 400 x 400mm with 200mm high collar, 100mm diameter	nr	0.80	8.09	14.55	3.40	26.03
Slates, size 400 x 400mm with 200mm high collar, 150mm diameter	nr	0.90	9.10	15.22	3.65	27.97

	Unit	Hours	Hours £	Materials £	O & P £	Total £
Copper sheet coverings						
Copper sheet 0.55mm thick in roof coverings to						
flat roof	m2	4.30	43.47	23.44	10.04	76.95
sloping roof	m2	4.50	45.50	23.44	10.34	79.28
Copper sheet 0.61mm thick in roof coverings to						
flat roof	m2	4.40	44.48	24.68	10.37	79.54
sloping roof	m2	4.60	46.51	24.68	10.68	81.86
Copper flashings, 0.55mm thick						
horizontal						
150mm girth	m	0.45	4.55	3.52	1.21	9.28
200mm girth	m	0.55	5.56	4.69	1.54	11.79
300mm girth	m	0.60	6.07	7.03	1.96	15.06
sloping						
150mm girth	m	0.55	5.56	3.52	1.36	10.44
200mm girth	m	0.60	6.07	4.69	1.61	12.37
300mm girth	m	0.65	6.57	7.03	2.04	15.64
Copper flashings, 0.61mm thick						
horizontal						
150mm girth	m	0.45	4.55	3.70	1.24	9.49
200mm girth	m	0.55	5.56	4.94	1.58	12.08
300mm girth	m	0.60	6.07	7.40	2.02	15.49
sloping						
150mm girth	m	0.55	5.56	3.70	1.39	10.65
200mm girth	m	0.60	6.07	4.94	1.65	12.66
300mm girth	m	0.65	6.57	7.40	2.10	16.07

	Unit	Hours	Hours £	Materials £	O & P £	Total £
Copper soakers size						
175 x 175mm	nr	0.25	2.53	1.20	0.56	4.29
300 x 175mm	nr	0.25	2.53	1.50	0.60	4.63

Built-up roofing

Built-up bituminous felt roof
coverings, the layers bonded with
hot bitumen

	Unit	Hours	Hours £	Materials £	O & P £	Total £
laid flat						
two layers	m2	0.35	3.54	4.88	1.26	9.68
three layers	m2	0.50	5.06	7.56	1.89	14.51
laid sloping						
two layers	m2	0.40	4.04	4.88	1.34	10.26
three layers	m2	0.55	5.56	7.56	1.97	15.09
Extra for						
aprons to two layer felt						
100mm girth	m	0.25	2.53	1.15	0.55	4.23
150mm girth	m	0.30	3.03	1.65	0.70	5.39
aprons to three layer felt						
100mm girth	m	0.25	2.53	1.35	0.58	4.46
150mm girth	m	0.30	3.03	1.80	0.72	5.56
skirtings to two layer felt						
100mm girth	m	0.25	2.53	1.15	0.55	4.23
150mm girth	m	0.30	3.03	1.65	0.70	5.39
skirtings to three layer felt						
100mm girth	m	0.25	2.53	1.35	0.58	4.46
150mm girth	m	0.30	3.03	1.80	0.72	5.56

	Unit	Hours	Hours £	Materials £	O & P £	Total £
flashings to two layer felt						
100mm girth	m	0.25	2.53	1.15	0.55	4.23
150mm girth	m	0.30	3.03	1.65	0.70	5.39
flashings to three layer felt						
100mm girth	m	0.25	2.53	1.35	0.58	4.46
150mm girth	m	0.30	3.03	1.80	0.72	5.56
Form collars around 100mm diameter pipe						
two layer	nr	1.25	12.64	1.87	2.18	16.68
three layer	nr	1.50	15.17	3.24	2.76	21.17
Form collars around 150mm diameter pipe						
two layer	nr	1.35	13.65	2.01	2.35	18.01
three layer	nr	1.60	16.18	3.65	2.97	22.80

	Unit	Specialist price £	O & P £	Total £

ASPHALT WORK

The following are specialist sub-contractor's prices

Damp proofing and tanking to BS 1097

20mm two coat mastic asphalt coverings to concrete surfaces

	Unit	Specialist price £	O & P £	Total £
flat	m2	27.15	4.07	31.22
sloping 10-45 degrees	m2	29.74	4.46	34.20
sloping 46-90 degrees	m2	32.47	4.87	37.34
vertical	m2	32.47	4.87	37.34

20mm two coat mastic asphalt coverings to concrete surfaces not exceeding 150mm wide

	Unit	Specialist price £	O & P £	Total £
flat	m	4.12	0.62	4.74
sloping 10-45 degrees	m	4.53	0.68	5.21
sloping 46-90 degrees	m	5.02	0.75	5.77
vertical	m	5.02	0.75	5.77

20mm two coat mastic asphalt coverings to concrete surfaces 150-300mm wide

	Unit	Specialist price £	O & P £	Total £
flat	m	8.15	1.22	9.37
sloping 10-45 degrees	m	8.98	1.35	10.33
sloping 46-90 degrees	m	9.80	1.47	11.27
vertical	m	9.80	1.47	11.27

30mm three coat mastic asphalt coverings to concrete surfaces

	Unit	Specialist price £	O & P £	Total £
flat	m2	48.71	7.31	56.02
sloping 10-45 degrees	m2	51.34	7.70	59.04

	Unit	Specialist price £	O & P £	Total £
sloping 46-90 degrees	m2	54.68	8.20	62.88
vertical	m2	54.68	8.20	62.88

30mm three coat mastic asphalt
coverings to concrete surfaces
not exceeding 150mm wide

flat	m	7.32	1.10	8.42
sloping 10-45 degrees	m	8.00	1.20	9.20
sloping 46-90 degrees	m	8.87	1.33	10.20
vertical	m	8.87	1.33	10.20

30mm three coat mastic asphalt
coverings to concrete surfaces
150-300mm wide

flat	m	14.69	2.20	16.89
sloping 10-45 degrees	m	16.33	2.45	18.78
sloping 46-90 degrees	m	17.80	2.67	20.47
vertical	m	17.80	2.67	20.47

Flooring to BS 1076

20mm two coat mastic asphalt
coverings to flat concrete surfaces

exceeding 300mm	m2	22.34	3.35	25.69
not exceeding 150mm wide	m	3.41	0.51	3.92
150-300mm wide	m	6.87	1.03	7.90

20mm two coat mastic asphalt
coverings to sloping concrete
surfaces

exceeding 300mm	m2	25.68	3.85	29.53
not exceeding 150mm wide	m	3.97	0.60	4.57
150-300mm wide	m	7.82	1.17	8.99

	Unit	Specialist price £	O & P £	Total £
Extra for				
working into recessed frames	m	1.98	0.30	2.28
working into channels 300mm girth	m	3.87	0.58	4.45
working into channels 600mm girth	m	7.88	1.18	9.06
30mm three coat mastic asphalt coverings to flat concrete surfaces				
exceeding 300mm	m2	41.57	6.24	47.81
not exceeding 150mm wide	m	6.32	0.95	7.27
150-300mm wide	m	12.67	1.90	14.57
30mm three coat mastic asphalt coverings to sloping concrete surfaces				
exceeding 300mm	m2	43.55	6.53	50.08
not exceeding 150mm wide	m	6.66	1.00	7.66
150-300mm wide	m	13.10	1.97	15.07
Extra for				
working into recessed frames	m	2.99	0.45	3.44
working into channels 300mm girth	m	6.98	1.05	8.03
working into channels 600mm girth	m	12.24	1.84	14.08

Roofing to BS 988

20mm two coat mastic asphalt coverings

	Unit	Specialist price £	O & P £	Total £
flat	m2	28.94	4.34	33.28
sloping 10-45 degrees	m2	30.56	4.58	35.14
sloping 46-90 degrees	m2	33.57	5.04	38.61
vertical	m2	33.57	5.04	38.61

	Unit	Specialist price £	O & P £	Total £
20mm two coat mastic asphalt coverings to concrete surfaces not exceeding 150mm wide				
flat	m	4.34	0.65	4.99
sloping 10-45 degrees	m	4.58	0.69	5.27
sloping 46-90 degrees	m	5.10	0.77	5.87
vertical	m	5.10	0.77	5.87
20mm two coat mastic asphalt coverings to concrete surfaces 150-300mm wide				
flat	m	8.76	1.31	10.07
sloping 10-45 degrees	m	9.20	1.38	10.58
sloping 46-90 degrees	m	10.10	1.52	11.62
vertical	m	10.10	1.52	11.62

	Unit	Hours	Hours £	Materials £	O & P £	Total £

FLOOR, WALL AND CEILING FINISHINGS

Screeds

Cement and sand (1:3) screed in beds with steel trowelled finish to level floors or to falls not exceeding 15 degrees from the horizontal

	Unit	Hours	Hours £	Materials £	O & P £	Total £
25mm thick						
over 300mm wide	m2	0.30	3.03	2.49	0.83	6.35
not exceeding 300mm wide	m	0.12	1.21	0.85	0.31	2.37
32mm thick						
over 300mm wide	m2	0.32	3.24	2.78	0.90	6.92
not exceeding 300mm wide	m	0.13	1.31	1.00	0.35	2.66
38mm thick						
over 300mm wide	m2	0.33	3.34	3.10	0.97	7.40
not exceeding 300mm wide	m	0.14	1.42	1.50	0.44	3.35
50mm thick						
over 300mm wide	m2	0.36	3.64	3.78	1.11	8.53
not exceeding 300mm wide	m	0.15	1.52	1.32	0.43	3.26
63mm thick						
over 300mm wide	m2	0.40	4.04	4.85	1.33	10.23
not exceeding 300mm wide	m	0.16	1.62	1.70	0.50	3.82
75mm thick						
over 300mm wide	m2	0.75	7.58	5.96	2.03	15.57
not exceeding 300mm wide	m	0.17	1.72	1.91	0.54	4.17

	Unit	Hours	Hours £	Materials £	O & P £	Total £
Cement and sand (1:3) screed in beds with steel trowelled finish to level landings						
25mm thick						
over 300mm wide	m2	0.40	4.04	2.49	0.98	7.51
not exceeding 300mm wide	m	0.15	1.52	0.85	0.35	2.72
32mm thick						
over 300mm wide	m2	0.42	4.25	2.78	1.05	8.08
not exceeding 300mm wide	m	0.16	1.62	1.00	0.39	3.01
38mm thick						
over 300mm wide	m2	0.43	4.35	3.10	1.12	8.56
not exceeding 300mm wide	m	0.17	1.72	1.50	0.48	3.70
50mm thick						
over 300mm wide	m2	0.46	4.65	3.78	1.26	9.70
not exceeding 300mm wide	m	0.18	1.82	1.32	0.47	3.61
63mm thick						
over 300mm wide	m2	0.50	5.06	4.85	1.49	11.39
not exceeding 300mm wide	m	0.19	1.92	1.70	0.54	4.16
75mm thick						
over 300mm wide	m2	0.85	8.59	5.96	2.18	16.74
not exceeding 300mm wide	m	0.33	3.34	1.91	0.79	6.03
Cement and sand (1:3) screed in beds with steel trowelled finish to risers and treads						
25mm thick						
275mm wide	m	0.11	1.11	0.85	0.29	2.26
350mm wide	m	0.13	1.31	0.95	0.34	2.60
32mm thick						
over 300mm wide	m2	0.13	1.31	0.95	0.34	2.60
not exceeding 300mm wide	m	0.15	1.52	1.05	0.38	2.95

	Unit	Hours	Hours £	Materials £	O & P £	Total £
38mm thick						
over 300mm wide	m2	0.16	1.62	1.45	0.46	3.53
not exceeding 300mm wide	m	0.18	1.82	1.55	0.51	3.88
50mm thick						
over 300mm wide	m2	0.17	1.72	1.30	0.45	3.47
not exceeding 300mm wide	m	0.19	1.92	1.40	0.50	3.82

Granolithic, cement and granite
chippings (1:2:5) with steel
trowelled finish to level floors or
to falls not exceeding 15 degrees
from the horizontal

	Unit	Hours	Hours £	Materials £	O & P £	Total £
25mm thick						
over 300mm wide	m2	0.40	4.04	3.60	1.15	8.79
not exceeding 300mm wide	m	0.14	1.42	1.12	0.38	2.92
32mm thick						
over 300mm wide	m2	0.42	4.25	3.90	1.22	9.37
not exceeding 300mm wide	m	0.15	1.52	1.20	0.41	3.12
38mm thick						
over 300mm wide	m2	0.43	4.35	4.22	1.29	9.85
not exceeding 300mm wide	m	0.16	1.62	1.32	0.44	3.38
50mm thick						
over 300mm wide	m2	0.46	4.65	4.88	1.43	10.96
not exceeding 300mm wide	m	0.17	1.72	1.51	0.48	3.71
63mm thick						
over 300mm wide	m2	0.50	5.06	6.02	1.66	12.74
not exceeding 300mm wide	m	0.18	1.82	1.85	0.55	4.22
75mm thick						
over 300mm wide	m2	0.85	8.59	7.21	2.37	18.17
not exceeding 300mm wide	m	0.22	2.22	2.20	0.66	5.09

	Unit	Hours	Hours £	Materials £	O & P £	Total £
Granolithic, cement and granite chippings (1:2:5) with steel trowelled finish to level landings						
25mm thick						
over 300mm wide	m2	0.50	5.06	3.60	1.30	9.95
not exceeding 300mm wide	m	0.16	1.62	1.12	0.41	3.15
32mm thick						
over 300mm wide	m2	0.52	5.26	3.90	1.37	10.53
not exceeding 300mm wide	m	0.17	1.72	1.20	0.44	3.36
38mm thick						
over 300mm wide	m2	0.53	5.36	4.22	1.44	11.02
not exceeding 300mm wide	m	0.18	1.82	1.32	0.47	3.61
50mm thick						
over 300mm wide	m2	0.56	5.66	4.88	1.58	12.12
not exceeding 300mm wide	m	0.19	1.92	1.51	0.51	3.95
Granolithic, cement and granite chippings (1:2:5) with steel trowelled finish to risers and treads						
25mm thick						
275mm wide	m	0.13	1.31	1.15	0.37	2.83
350mm wide	m	0.15	1.52	1.25	0.41	3.18
32mm thick						
over 300mm wide	m2	0.15	1.52	1.25	0.41	3.18
not exceeding 300mm wide	m	0.17	1.72	1.35	0.46	3.53
38mm thick						
over 300mm wide	m2	0.18	1.82	1.80	0.54	4.16
not exceeding 300mm wide	m	0.20	2.02	1.90	0.59	4.51

	Unit	Hours	Hours £	Materials £	O & P £	Total £

In situ wall coverings

Cement and sand (1:3) beds and
backings to brickwork or blockwork walls

	Unit	Hours	Hours £	Materials £	O & P £	Total £
13mm thick						
over 300mm wide	m2	0.25	2.53	1.05	0.54	4.11
not exceeding 300mm wide	m	0.10	1.01	0.38	0.21	1.60
19mm thick						
over 300mm wide	m2	0.28	2.83	1.88	0.71	5.42
not exceeding 300mm wide	m	0.12	1.21	0.67	0.28	2.17

Cement and sand (1:3) beds and
backings to sides of isolated
columns

	Unit	Hours	Hours £	Materials £	O & P £	Total £
13mm thick						
over 300mm wide	m2	0.35	3.54	1.05	0.69	5.28
not exceeding 300mm wide	m	0.13	1.31	0.38	0.25	1.95
19mm thick						
over 300mm wide	m2	0.38	3.84	1.88	0.86	6.58
not exceeding 300mm wide	m	0.15	1.52	0.67	0.33	2.51

Premixed lightweight plaster,
11mm bonding, 2mm finish

	Unit	Hours	Hours £	Materials £	O & P £	Total £
brickwork or blockwork walls						
over 300mm wide	m2	0.48	4.85	2.04	1.03	7.93
not exceeding 300mm wide	m	0.18	1.82	0.70	0.38	2.90
isolated columns						
over 300mm wide	m2	0.58	5.86	2.04	1.19	9.09
not exceeding 300mm wide	m	0.22	2.22	0.70	0.44	3.36
ceilings						
over 300mm wide	m2	0.62	6.27	1.88	1.22	9.37
not exceeding 300mm wide	m	0.24	2.43	0.67	0.46	3.56

	Unit	Hours	Hours £	Materials £	O & P £	Total £
One coat 'Universal' plaster to brickwork or blockwork walls						
13mm thick						
over 300mm wide	m2	0.30	3.03	2.61	0.85	6.49
not exceeding 300mm wide	m	0.11	1.11	0.85	0.29	2.26
19mm thick						
over 300mm wide	m2	0.35	3.54	3.23	1.02	7.78
not exceeding 300mm wide	m	0.13	1.31	1.02	0.35	2.68
One coat 'Universal' plaster to isolated columns						
13mm thick						
over 300mm wide	m2	0.40	4.04	2.61	1.00	7.65
not exceeding 300mm wide	m	0.14	1.42	0.85	0.34	2.61
19mm thick						
over 300mm wide	m2	0.45	4.55	3.23	1.17	8.95
not exceeding 300mm wide	m	0.16	1.62	1.02	0.40	3.03
One coat board finish 3mm thick						
plasterboard walls						
over 300mm wide	m2	0.26	2.63	1.22	0.58	4.43
not exceeding 300mm wide	m	0.10	1.01	0.46	0.22	1.69
plasterboard ceilings						
over 300mm wide	m2	0.36	3.64	1.22	0.73	5.59
not exceeding 300mm wide	m	0.14	1.42	0.46	0.28	2.16
plasterboard isolated columns						
over 300mm wide	m2	0.36	3.64	1.22	0.73	5.59
not exceeding 300mm wide	m	0.14	1.42	0.46	0.28	2.16

	Unit	Hours	Hours £	Materials £	O & P £	Total £
Three coat Tyrolean rendering 22mm thick to brickwork or blockwork walls						
over 300mm wide	m2	0.85	8.59	4.22	1.92	14.74
not exceeding 300mm wide	m	0.32	3.24	1.48	0.71	5.42
Lathing						
Expamet expanded metal lathing, 6mm thick						
stapled to softwood, vertically						
over 300mm wide	m2	0.64	6.47	4.67	1.67	12.81
not exceeding 300mm wide	m	0.22	2.22	1.40	0.54	4.17
tied with wire, vertically						
over 300mm wide	m2	0.66	6.67	4.67	1.70	13.04
not exceeding 300mm wide	m	0.24	2.43	1.40	0.57	4.40
tied with wire, horizontally						
over 300mm wide	m2	0.74	7.48	4.67	1.82	13.97
not exceeding 300mm wide	m	0.26	2.63	1.40	0.60	4.63
Expamet Riblath expanded metal lathing, 0.3mm thick						
stapled to softwood, vertically						
over 300mm wide	m2	0.54	5.46	5.62	1.66	12.74
not exceeding 300mm wide	m	0.18	1.82	1.84	0.55	4.21
tied with wire, vertically						
over 300mm wide	m2	0.56	5.66	5.62	1.69	12.97
not exceeding 300mm wide	m	0.20	2.02	1.84	0.58	4.44
tied with wire, horizontally						
over 300mm wide	m2	0.64	6.47	5.62	1.81	13.90
not exceeding 300mm wide	m	0.24	2.43	1.84	0.64	4.91

	Unit	Hours	Hours £	Materials £	O & P £	Total £

Plasterboard

Tapered edge wallboard, open
joints to receive skimming or
similar, over 300mm wide

	Unit	Hours	Hours £	Materials £	O & P £	Total £
9.5mm thick						
to ceilings over 300mm wide	m2	0.35	3.54	1.96	0.82	6.32
to beams not exceeding						
600mm	m	0.22	2.22	1.05	0.49	3.77
12.5mm thick						
to ceilings over 300mm wide	m2	0.40	4.04	2.36	0.96	7.36
to beams not exceeding						
600mm	m	0.25	2.53	1.24	0.57	4.33

Floor tiling

Red clay quarry tiles, bedded in
cement mortar (1:3) 12mm thick,
butt jointed straight both ways

	Unit	Hours	Hours £	Materials £	O & P £	Total £
150 x 150 x 12.5mm thick to floors						
over 300mm wide	m2	0.90	9.10	24.88	5.10	39.08
not exceeding 300mm wide	m	0.36	3.64	7.62	1.69	12.95
200 x 200 x 12.5mm thick to floors						
over 300mm wide	m2	0.80	8.09	41.33	7.41	56.83
not exceeding 300mm wide	m	0.32	3.24	14.67	2.69	20.59
150 x 150 x 12.5mm thick to landings						
over 300mm wide	m2	1.00	10.11	24.88	5.25	40.24
not exceeding 300mm wide	m	0.45	4.55	7.62	1.83	13.99

	Unit	Hours	Hours £	Materials £	O & P £	Total £
200 x 200 x 12.5mm thick to landings						
over 300mm wide	m2	0.90	9.10	41.33	7.56	57.99
not exceeding 300mm wide	m	0.36	3.64	14.67	2.75	21.06
Brown clay quarry tiles, bedded in cement mortar (1:3) 12mm thick, butt jointed straight both ways						
150 x 150 x 12.5mm thick to floors						
over 300mm wide	m2	0.90	9.10	29.68	5.82	44.60
not exceeding 300mm wide	m	0.36	3.64	10.27	2.09	16.00
200 x 200 x 12.5mm thick to floors						
over 300mm wide	m2	0.80	8.09	43.82	7.79	59.69
not exceeding 300mm wide	m	0.32	3.24	14.71	2.69	20.64
150 x 150 x 12.5mm thick to landings						
over 300mm wide	m2	1.00	10.11	29.68	5.97	45.76
not exceeding 300mm wide	m	0.49	4.95	10.27	2.28	17.51
200 x 200 x 12.5mm thick to landings						
over 300mm wide	m2	0.90	9.10	43.82	7.94	60.86
not exceeding 300mm wide	m	0.36	3.64	14.71	2.75	21.10
Vinyl floor tiling size 300 x 300mm fixed with adhesive						
2mm thick						
over 300mm wide	m2	0.28	2.83	6.98	1.47	11.28
not exceeding 300mm wide	m	0.10	1.01	2.15	0.47	3.64
2.5mm thick						
over 300mm wide	m2	0.30	3.03	8.02	1.66	12.71
not exceeding 300mm wide	m	0.11	1.11	2.40	0.53	4.04

	Unit	Hours	Hours £	Materials £	O & P £	Total £
3mm thick						
over 300mm wide	m2	0.32	3.24	8.02	1.69	12.94
not exceeding 300mm wide	m	0.12	1.21	2.40	0.54	4.16
Vinyl floor sheeting with welded joints, fixed with adhesive						
2mm thick						
over 300mm wide	m2	0.36	3.64	8.44	1.81	13.89
not exceeding 300mm wide	m	0.14	1.42	3.12	0.68	5.22
2.5mm thick						
over 300mm wide	m2	0.40	4.04	11.47	2.33	17.84
not exceeding 300mm wide	m	0.18	1.82	4.12	0.89	6.83
Linoleum floor sheeting with butt joints, fixed to adhesive						
2.5mm thick						
over 300mm wide	m2	0.34	3.44	8.79	1.83	14.06
not exceeding 300mm wide	m	0.13	1.31	3.54	0.73	5.58
3.2mm thick						
over 300mm wide	m2	0.36	3.64	9.64	1.99	15.27
not exceeding 300mm wide	m	0.14	1.42	3.57	0.75	5.73

Wall tiling

White glazed ceramic wall tiling, fixed with adhesive, pointing with white grout

108 x 108 x 4mm thick

	Unit	Hours	Hours £	Materials £	O & P £	Total £
over 300mm wide	m2	0.95	9.60	17.88	4.12	31.61
not exceeding 300mm wide	m	0.38	3.84	6.20	1.51	11.55

	Unit	Hours	Hours £	Materials £	O & P £	Total £
152 x 152 x 5.6mm thick						
over 300mm wide	m2	0.85	8.59	16.74	3.80	29.13
not exceeding 300mm wide	m	0.34	3.44	5.76	1.38	10.58
203 x 102 x 6.5mm thick						
over 300mm wide	m2	0.75	7.58	15.77	3.50	26.86
not exceeding 300mm wide	m	0.30	3.03	5.55	1.29	9.87
Extra for						
cutting to edges	m	0.10	1.01	0.00	0.15	1.16
holes for small pipes	nr	0.20	2.02	0.00	0.30	2.33
holes for large pipes	nr	0.30	3.03	0.00	0.45	3.49

Dark-coloured glazed ceramic wall
tiling, fixed with adhesive, pointing
with matching grout

	Unit	Hours	Hours £	Materials £	O & P £	Total £
108 x 108 x 4mm thick						
over 300mm wide	m2	0.95	9.60	22.34	4.79	36.74
not exceeding 300mm wide	m	0.38	3.84	7.54	1.71	13.09
152 x 152 x 5.6mm thick						
over 300mm wide	m2	0.85	8.59	21.23	4.47	34.30
not exceeding 300mm wide	m	0.34	3.44	7.00	1.57	12.00
203 x 102 x 6.5mm thick						
over 300mm wide	m2	0.75	7.58	20.11	4.15	31.85
not exceeding 300mm wide	m	0.30	3.03	6.55	1.44	11.02
Extra for						
cutting to edges	m	0.10	1.01	0.00	0.15	1.16
holes for small pipes	nr	0.20	2.02	0.00	0.30	2.33
holes for large pipes	nr	0.30	3.03	0.00	0.45	3.49

	Unit	Hours	Hours £	Materials £	O & P £	Total £

PLUMBING AND HEATING

Sanitary fittings

Acrylic reinforced bath, size 1700
x 750mm, complete with 2nr
chromium plated grips, 40mm
waste fitting, overflow with chain
and plastic plug

	Unit	Hours	Hours £	Materials £	O & P £	Total £
white	nr	3.00	32.82	164.40	29.58	226.80
coloured	nr	3.00	32.82	164.40	29.58	226.80

Porcelain enamel standard gauge
bath, size 1700 x 700mm, complete
with 2nr chromium plated grips,
40mm waste fitting, overflow
with chain and plastic plug

	Unit	Hours	Hours £	Materials £	O & P £	Total £
white	nr	3.00	32.82	112.00	21.72	166.54
coloured	nr	3.00	32.82	116.23	22.36	171.41

Bath panels, moulded acrylic,
fixing in position for trimming as
as required

	Unit	Hours	Hours £	Materials £	O & P £	Total £
end panel	nr	0.30	3.28	7.64	1.64	12.56
side panel	nr	0.50	5.47	13.98	2.92	22.37

Angle strip polished aluminium,
fixing with chromium plated dome
headed screws, cutting to length

	Unit	Hours	Hours £	Materials £	O & P £	Total £
25 x 25 x 560mm long	nr	0.35	3.83	2.02	0.88	6.73

Wash basin, vitreous china, size
510 x 410mm, complete with
pedestal, chromium plated waste,
overflow with chain and plastic
plug

	Unit	Hours	Hours £	Materials £	O & P £	Total £
white	nr	2.20	24.07	71.76	14.37	110.20
coloured	nr	2.20	24.07	79.24	15.50	118.80

	Unit	Hours	Hours £	Materials £	O & P £	Total £
Wash basin, vitreous china, size 610 x 470mm, complete with brackets, chromium plated waste, overflow with chain and plastic plug						
white	nr	2.30	25.16	55.06	12.03	92.26
coloured	nr	2.30	25.16	55.06	12.03	92.26
Stainless steel single roll-edge type sink, complete with inset chromium plated waste, overflow with chain and plastic plug, size						
1000 x 500mm	nr	1.30	14.22	79.43	14.05	107.70
1000 x 600mm	nr	1.40	15.32	86.68	15.30	117.30
1200 x 600mm	nr	1.50	16.41	99.08	17.32	132.81
Stainless steel double sit-on type sink, complete with inset chromium plated waste, overflow with chain and plastic plug, size						
1500 x 500mm	nr	1.80	19.69	97.03	17.51	134.23
1500 x 600mm	nr	1.90	20.79	113.73	20.18	154.69
Vitreous china low level WC suite comprising pan, plastic seat and cover, 9 litre cistern, low pressure ball valve, connecting pipework screwed to floor	nr	2.35	25.71	192.58	32.74	251.03
Vitreous china low level close coupled WC suite comprising pan, plastic seat and cover, 9 litre cistern, low pressure ball valve, connecting pipework screwed to floor	nr	2.45	26.80	243.82	40.59	311.22
Free-standing plain rim, vitreous china bidet excluding fittings						
white	nr	2.50	27.35	231.09	38.77	297.21
coloured	nr	2.50	27.35	250.59	41.69	319.63

	Unit	Hours	Hours £	Materials £	O & P £	Total £
Bowl urinals, white vitreous china, wall-mounted on hangars chromium plated dome outlet						
430mm wide x 305mm high	nr	1.00	10.94	97.48	16.26	124.68
500mm wide x 350mm high	nr	1.10	12.03	170.42	27.37	209.82
Chromium plated pillar taps						
13mm	pr	0.40	4.38	23.88	4.24	32.49
19mm	pr	0.45	4.92	31.92	5.53	42.37
Chromium plated mixer taps, 19mm						
cross-top handles	pr	0.60	6.56	47.67	8.14	62.37
lever handles	pr	0.60	6.56	70.05	11.49	88.11
Chromium plated thermostatic exposed shower valve with flexible hose, slide rail and spray set	nr	1.40	15.32	182.64	29.69	227.65
Instant electric 9.5kw shower with flexible hose, slide rail and spray set	nr	1.50	16.41	137.88	23.14	177.43
Shower cubicle size 788 x 842 x 2115mm in anodised aluminium frame and safety glass	nr	2.00	21.88	666.37	103.24	791.49
White acrylic fireclay shower tray size 750 x 750 x 180mm with chromium plated waste fitting	nr	1.30	14.22	87.17	15.21	116.60

Holes and chases

	Unit	Hours	Hours £	Materials £	O & P £	Total £
Cutting holes for pipes up to 25mm diameter in walls 102.5mm thick						
commons	nr	0.40	4.38	0.00	0.66	5.03
facings	nr	0.70	7.66	0.00	1.15	8.81
engineering class A	nr	0.90	9.85	0.00	1.48	11.32

	Unit	Hours	Hours £	Materials £	O & P £	Total £
Cutting holes for pipes up to 25mm diameter in walls 100mm thick						
concrete blocks	nr	0.30	3.28	0.00	0.49	3.77
Thermalite blocks	nr	0.20	2.19	0.00	0.33	2.52
aerated concrete blocks	nr	0.25	2.74	0.00	0.41	3.15
Cutting and pinning ends of pipe support brackets and make good to						
concrete	nr	0.45	4.92	0.00	0.74	5.66
brickwork	nr	0.35	3.83	0.00	0.57	4.40
blockwork	nr	0.30	3.28	0.00	0.49	3.77
tiled walls	nr	0.50	5.47	0.00	0.82	6.29
Cut opening through cavity wall comprising facing bricks and concrete blocks						
balanced flue outlet	nr	0.90	9.85	0.00	1.48	11.32
150mm diameter flue pipe	nr	0.75	8.21	0.00	1.23	9.44
19mm overflow pipe	nr	0.20	2.19	0.00	0.33	2.52
Cutting chases for one pipe up to 25mm diameter in						
commons	nr	0.35	3.83	0.00	0.57	4.40
facings	nr	0.38	4.16	0.00	0.62	4.78
engineering class A	nr	0.45	4.92	0.00	0.74	5.66
engineering class B	nr	0.40	4.38	0.00	0.66	5.03
concrete blocks	nr	0.35	3.83	0.00	0.57	4.40
Thermalite blocks	nr	0.25	2.74	0.00	0.41	3.15
aerated concrete blocks	nr	0.20	2.19	0.00	0.33	2.52
plastered surfaces	m	0.15	1.64	0.50	0.32	2.46
tiled surfaces	m	0.25	2.74	2.00	0.71	5.45
concrete floor	m	0.35	3.83	0.50	0.65	4.98
granolithic floor	m	0.50	5.47	0.60	0.91	6.98
concrete soffit	m	0.55	6.02	0.50	0.98	7.49

	Unit	Hours	Hours £	Materials £	O & P £	Total £

Rainwater pipes and gutters

PVC-U rainwater pipe plugged
to brickwork with pipe brackets
and fitting clips at 2m maximum
centres

	Unit	Hours	Hours £	Materials £	O & P £	Total £
68mm diameter pipe	m	0.25	2.74	4.30	1.06	8.09
extra over for						
bend, 87.5 degrees	nr	0.25	2.74	2.94	0.85	6.53
offset	nr	0.25	2.74	6.85	1.44	11.02
branch	nr	0.25	2.74	7.00	1.46	11.20
shoe	nr	0.25	2.74	2.55	0.79	6.08
access pipe	nr	0.25	2.74	6.33	1.36	10.42
hopper head	nr	0.25	2.74	10.56	1.99	15.29
68mm square pipe	m	0.25	2.74	4.72	1.12	8.57
extra over for						
bend, 87.5 degrees	nr	0.25	2.74	3.02	0.86	6.62
offset	nr	0.25	2.74	6.98	1.46	11.17
branch	nr	0.25	2.74	7.36	1.51	11.61
shoe	nr	0.25	2.74	2.94	0.85	6.53
access pipe	nr	0.25	2.74	7.22	1.49	11.45
hopper head	nr	0.25	2.74	11.23	2.09	16.06

Cast iron rainwater pipe,
straight, ears cast on, spigot
and socket dry joints plugged
to brickwork

	Unit	Hours	Hours £	Materials £	O & P £	Total £
75mm diameter pipe	m	0.30	3.28	17.28	3.08	23.65
extra over for						
offset, 150mm	nr	0.30	3.28	12.42	2.36	18.06
offset, 229mm	nr	0.30	3.28	14.88	2.72	20.89
offset, 305mm	nr	0.30	3.28	17.23	3.08	23.59
bend	nr	0.30	3.28	5.90	1.38	10.56
branch	nr	0.30	3.28	6.47	1.46	11.21
eared shoe	nr	0.30	3.28	13.96	2.59	19.83
flat hopper	nr	0.30	3.28	10.88	2.12	16.29
rectangular hopper	nr	0.30	3.28	23.46	4.01	30.75

	Unit	Hours	Hours £	Materials £	O & P £	Total £
100mm diameter pipe	m	0.35	3.83	23.98	4.17	31.98
extra over for						
offset, 150mm	nr	0.35	3.83	22.66	3.97	30.46
offset, 229mm	nr	0.35	3.83	27.44	4.69	35.96
offset, 305mm	nr	0.35	3.83	27.44	4.69	35.96
bend	nr	0.35	3.83	19.82	3.55	27.20
branch	nr	0.35	3.83	7.92	1.76	13.51
eared shoe	nr	0.35	3.83	16.70	3.08	23.61
flat hopper	nr	0.35	3.83	16.70	3.08	23.61
rectangular hopper	nr	0.35	3.83	25.20	4.35	33.38
PVC-U half round rainwater gutter, fixed to timber with support brackets at 1m maximum centres						
76mm gutter	m	0.20	2.19	3.88	0.91	6.98
extra over for						
running outlet	nr	0.20	2.19	2.22	0.66	5.07
angle	nr	0.20	2.19	2.64	0.72	5.55
stop end outlet	nr	0.10	1.09	2.89	0.60	4.58
stop end	nr	0.10	1.09	1.45	0.38	2.93
112mm gutter	m	0.26	2.84	4.75	1.14	8.73
extra over for						
running outlet	nr	0.26	2.84	2.53	0.81	6.18
angle	nr	0.26	2.84	2.99	0.88	6.71
stop end outlet	nr	0.13	1.42	3.12	0.68	5.22
stop end	nr	0.13	1.42	1.76	0.48	3.66
Cast iron half round rainwater gutter with mastic joints, primed, fixed to timber with support brackets at 1m maximum centres						
100mm gutter	m	0.36	3.94	11.76	2.35	18.05
extra over for						
running outlet	nr	0.36	3.94	6.54	1.57	12.05
angle	nr	0.36	3.94	6.70	1.60	12.23

	Unit	Hours	Hours £	Materials £	O & P £	Total £
Plastic wire balloon guard for pipes and outlets, diameter						
50mm	nr	0.05	0.55	1.87	0.36	2.78
63mm	nr	0.05	0.55	1.92	0.37	2.84
75mm	nr	0.05	0.55	1.95	0.37	2.87
100mm	nr	0.05	0.55	2.02	0.39	2.95

Waste pipes

	Unit	Hours	Hours £	Materials £	O & P £	Total £
MPVC-U waste system, solvent welded joints, clips at 500mm maximum centres, plugged to brickwork						
32mm diameter pipe	m	0.25	2.74	2.38	0.77	5.88
extra over for						
bend, 45 degrees	nr	0.24	2.63	1.05	0.55	4.23
bend, 87.5 degrees	nr	0.24	2.63	1.22	0.58	4.42
long tail bend, 90 degrees	nr	0.24	2.63	1.29	0.59	4.50
tee	nr	0.24	2.63	1.73	0.65	5.01
tank connector	nr	0.24	2.63	2.17	0.72	5.51
universal connector	nr	0.24	2.63	1.67	0.64	4.94
bottle trap	nr	0.24	2.63	3.74	0.95	7.32
tubular P trap	nr	0.24	2.63	4.00	0.99	7.62
tubular S trap	nr	0.24	2.63	4.38	1.05	8.06
running P trap	nr	0.24	2.63	5.17	1.17	8.96
connection to back inlet gulley, caulking bush	nr	0.15	1.64	2.87	0.68	5.19
50mm diameter pipe	m	0.34	3.72	2.88	0.99	7.59
extra over for						
bend, 45 degrees	nr	0.32	3.50	1.90	0.81	6.21
bend, 87.5 degrees	nr	0.32	3.50	2.21	0.86	6.57
long tail bend, 90 degrees	nr	0.32	3.50	2.42	0.89	6.81
tee	nr	0.32	3.50	3.16	1.00	7.66
cross tee	nr	0.32	3.50	5.62	1.37	10.49

	Unit	Hours	Hours £	Materials £	O & P £	Total £
MPVC-U waste system, push-fit joints, clips at 500mm maximum centres, plugged to brickwork						
32mm diameter pipe	m	0.20	2.19	1.18	0.51	3.87
extra over for						
bend, 45 degrees	nr	0.18	1.97	0.78	0.41	3.16
knuckle bend, 90 degrees	nr	0.18	1.97	0.78	0.41	3.16
spigot bend, 90 degrees	nr	0.18	1.97	0.78	0.41	3.16
tee	nr	0.18	1.97	0.78	0.41	3.16
P trap, 38mm	nr	0.18	1.97	2.40	0.66	5.02
P trap, 76mm	nr	0.18	1.97	3.00	0.75	5.71
bottle trap, 38mm	nr	0.18	1.97	2.70	0.70	5.37
bottle trap, 76mm	nr	0.18	1.97	2.60	0.69	5.25
anti-syphon bottle trap, 76mm	nr	0.18	1.97	3.95	0.89	6.81
connection to back inlet						
gulley, caulking bush	nr	0.16	1.75	2.87	0.69	5.31
50mm diameter pipe	m	0.28	3.06	1.97	0.75	5.79
extra over for						
socket reducer	nr	0.26	2.84	1.40	0.64	4.88
bend, 45 degrees	nr	0.26	2.84	1.40	0.64	4.88
knuckle bend, 90 degrees	nr	0.26	2.84	1.40	0.64	4.88
spigot bend, 90 degrees	nr	0.26	2.84	1.40	0.64	4.88
tee	nr	0.26	2.84	1.40	0.64	4.88

Soil pipes

	Unit	Hours	Hours £	Materials £	O & P £	Total £
PVC-U soil system, solvent-welded joints, holderbats at 1250mm maximum centres, plugged to brickwork						
110mm diameter pipe	m	0.38	4.16	8.04	1.83	14.03
extra over for						
bend, 92.5 degrees	nr	0.34	3.72	9.20	1.94	14.86
bend, 135 degrees	nr	0.34	3.72	19.71	3.51	26.94
bend, variable	nr	0.34	3.72	17.58	3.19	24.49
bend, spigot/spigot	nr	0.34	3.72	9.06	1.92	14.70

	Unit	Hours	Hours £	Materials £	O & P £	Total £
bend, spigot/socket	nr	0.34	3.72	10.45	2.13	16.30
branch, single, 92.5 degrees	nr	0.40	4.38	11.96	2.45	18.79
branch, single, 135 degrees	nr	0.40	4.38	14.51	2.83	21.72
branch, single, 92.5 degrees spigot outlet	nr	0.40	4.38	16.04	3.06	23.48
branch, single, 92.5 degrees socket outlet	nr	0.40	4.38	27.02	4.71	36.11
branch, double, 92.5 degrees	nr	0.40	4.38	33.83	5.73	43.94
branch, double, 135 degrees	nr	0.40	4.38	35.37	5.96	45.71
branch, corner, 92.5 degrees spigot outlet	nr	0.40	4.38	61.22	9.84	75.44
branch, corner, 92.5 degrees socket outlet	nr	0.40	4.38	61.22	9.84	75.44
access coupling	nr	0.34	3.72	21.97	3.85	29.54
access pipe connector	nr	0.34	3.72	19.80	3.53	27.05
access pipe, single branch	nr	0.34	3.72	28.10	4.77	36.59
access door	nr	0.38	4.16	10.88	2.26	17.29
access cap	nr	0.38	4.16	10.88	2.26	17.29
WC manifold connector	nr	0.38	4.16	12.10	2.44	18.70
vent cowl	nr	0.22	2.41	2.27	0.70	5.38

Cast iron soil system with flexible joints, pipe brackets at 2m maximum centres, plugged to brickwork

	Unit	Hours	Hours £	Materials £	O & P £	Total £
100mm diameter pipe	m	0.65	7.11	29.47	5.49	42.07
extra over for						
bend, short radius	nr	0.55	6.02	13.62	2.95	22.58
bend, short radius with door	nr	0.55	6.02	28.80	5.22	40.04
bend, long radius	nr	0.55	6.02	22.06	4.21	32.29
bend, long radius, access	nr	0.55	6.02	37.26	6.49	49.77
bend, long radius, heel rest	nr	0.55	6.02	26.71	4.91	37.64
bend, long tail, 87.5 degrees	nr	0.55	6.02	17.60	3.54	27.16
branch, single, plain	nr	0.60	6.56	21.05	4.14	31.76
branch, single with door	nr	0.60	6.56	36.27	6.43	49.26
branch, double, plain	nr	0.60	6.56	26.05	4.89	37.51
branch, double with door	nr	0.60	6.56	41.24	7.17	54.97
branch, corner	nr	0.60	6.56	34.27	6.13	46.96
offset, 75mm projection	nr	0.55	6.02	14.33	3.05	23.40

	Unit	Hours	Hours £	Materials £	O & P £	Total £
offset, 115mm projection	nr	0.55	6.02	17.08	3.46	26.56
offset, 150mm projection	nr	0.55	6.02	17.08	3.46	26.56
offset, 225mm projection	nr	0.55	6.02	19.58	3.84	29.44
offset, 300mm projection	nr	0.55	6.02	22.06	4.21	32.29
blank end, plain	nr	0.55	6.02	4.15	1.53	11.69
blank end, drilled and tapped	nr	0.55	6.02	9.02	2.26	17.29
P trap, plain	nr	0.55	6.02	21.83	4.18	32.02
P trap with door	nr	0.55	6.02	37.03	6.46	49.50
WC connector, 305mm effective length	nr	0.55	6.02	12.39	2.76	21.17

Overflows

MPVC-U overflow system, solvent-welded joints, clips at 500mm maximum centres, plugged to brickwork

19mm diameter pipe	m	0.20	2.19	1.42	0.54	4.15
extra over for						
bend	nr	0.18	1.97	1.73	0.55	4.25
tee	nr	0.18	1.97	2.52	0.67	5.16
connector, bent tank	nr	0.20	2.19	1.31	0.52	4.02

Traps

Polypropylene traps, screwed joints to outlet and pipe

Bottle trap, 76mm seal						
32mm	nr	0.30	3.28	3.89	1.08	8.25
40mm	nr	0.35	3.83	4.92	1.31	10.06
Anti-syphon bottle trap						
32mm	nr	0.30	3.28	4.65	1.19	9.12
40mm	nr	0.35	3.83	5.30	1.37	10.50
Tubular S trap						
32mm	nr	0.30	3.28	4.83	1.22	9.33
40mm	nr	0.35	3.83	6.81	1.60	12.23

	Unit	Hours	Hours £	Materials £	O & P £	Total £
32mm	nr	0.30	3.28	5.13	1.26	9.67
40mm	nr	0.35	3.83	4.63	1.27	9.73
Running P trap						
32mm	nr	0.30	3.28	5.93	1.38	10.59
40mm	nr	0.35	3.83	6.44	1.54	11.81
Bath trap						
40mm	nr	0.35	3.83	4.98	1.32	10.13
Bath trap with overflow						
40mm	nr	0.35	3.83	11.50	2.30	17.63

Copper pipework, capillary joints

Copper pipe to BS 2871 Table X, lead-free pre-soldered capillary joints and fittings to BS 64, clips at 1250mm maximum centres

	Unit	Hours	Hours £	Materials £	O & P £	Total £
15mm diameter, to timber	m	0.20	2.19	1.60	0.57	4.36
15mm diameter, plugged and screwed	m	0.22	2.41	1.88	0.64	4.93
extra over for						
straight coupling	nr	0.18	1.97	0.19	0.32	2.48
reduced coupling, 15 x 8mm	nr	0.18	1.97	1.66	0.54	4.17
reduced coupling, 15 x 10mm	nr	0.18	1.97	1.50	0.52	3.99
reduced coupling, 15 x 12mm	nr	0.18	1.97	1.45	0.51	3.93
adaptor coupling, 15mm x 1/2in	nr	0.18	1.97	1.86	0.57	4.40
straight female connector, 15mm x 1/2in	nr	0.18	1.97	2.24	0.63	4.84
female reducing connector, 15mm x 3/8in	nr	0.18	1.97	4.37	0.95	7.29
female reducing connector, 15mm x 3/4in	nr	0.18	1.97	5.45	1.11	8.53
straight male connector, 15mm x 1/2in	nr	0.18	1.97	1.91	0.58	4.46

	Unit	Hours	Hours £	Materials £	O & P £	Total £
male reducing connector, 15mm x 3/4in	nr	0.18	1.97	4.89	1.03	7.89
male reducing connector, 15mm x 3/4in	nr	0.18	1.97	3.42	0.81	6.20
lead connector	nr	0.18	1.97	1.62	0.54	4.13
tank connector, 15mm x 1/2in	nr	0.18	1.97	4.55	0.98	7.50
tank connector, 15mm x 3/4in	nr	0.18	1.97	10.43	1.86	14.26
reducer, 15 x 8mm	nr	0.18	1.97	1.47	0.52	3.96
reducer, 15 x 10mm	nr	0.18	1.97	0.79	0.41	3.17
reducer, 15 x 12mm	nr	0.18	1.97	1.30	0.49	3.76
female adaptor, 15 x 1/2in	nr	0.18	1.97	3.78	0.86	6.61
male adaptor, 15 x 1/2in	nr	0.18	1.97	3.86	0.87	6.70
adaptor, 15 x 1/2in	nr	0.18	1.97	1.60	0.54	4.10
elbow, 15mm	nr	0.18	1.97	0.45	0.36	2.78
male elbow, 15mm x 1/2in	nr	0.18	1.97	3.74	0.86	6.57
reducing male elbow, 15mm x 3/8in	nr	0.18	1.97	6.05	1.20	9.22
female elbow, 15mm x 1/2in	nr	0.18	1.97	3.28	0.79	6.04
reducing female elbow, 15mm x 1/4in	nr	0.18	1.97	5.14	1.07	8.18
backplate elbow, 15mm x 1/2in	nr	0.18	1.97	4.15	0.92	7.04
flanged bend, 15mm x 1/2in	nr	0.18	1.97	8.44	1.56	11.97
flanged bend, 15mm x 3/4in	nr	0.18	1.97	8.79	1.61	12.37
overflow bend, 15mm x 1/2in	nr	0.18	1.97	9.80	1.77	13.53
slow bend	nr	0.18	1.97	0.67	0.40	3.04
return bend	nr	0.18	1.97	4.99	1.04	8.00
obtuse elbow	nr	0.18	1.97	0.85	0.42	3.24
tee, reduced branch largest end 15mm	nr	0.22	2.41	2.36	0.72	5.48
tee, one end and one branch reduced, largest end, 15mm	nr	0.22	2.41	4.89	1.09	8.39
tee, both ends reduced, largest end 15mm	nr	0.22	2.41	4.37	1.02	7.79
equal tee, 15mm	nr	0.22	2.41	1.20	0.54	4.15
sweep tee, 90 degrees, 15mm	nr	0.22	2.41	6.60	1.35	10.36
offset tee, 15mm	nr	0.22	2.41	8.75	1.67	12.83
double sweep tee, 15mm	nr	0.22	2.41	7.48	1.48	11.37
cross equal tee, 15mm	nr	0.22	2.41	9.91	1.85	14.16
stop end, 15mm	nr	0.18	1.97	1.04	0.45	3.46

	Unit	Hours	Hours £	Materials £	O & P £	Total £
tap connector, 15mm	nr	0.18	1.97	2.10	0.61	4.68
bent union adaptor, 15mm x 3/4in	nr	0.18	1.97	4.56	0.98	7.51
bent male union adaptor, 15mm x 1/2in	nr	0.18	1.97	7.62	1.44	11.03
bent female union adaptor, 15mm x 1/2in	nr	0.18	1.97	7.62	1.44	11.03
made bend	nr	0.18	1.97	0.00	0.30	2.26
22mm diameter, to timber	m	0.20	2.19	3.13	0.80	6.12
22mm diameter, plugged and screwed	m	0.22	2.41	3.39	0.87	6.67
extra over for						
straight coupling	nr	0.20	2.19	0.32	0.38	2.88
reduced coupling, 22 x 8mm	nr	0.20	2.19	2.58	0.72	5.48
reduced coupling, 22 x 10mm	nr	0.20	2.19	2.58	0.72	5.48
reduced coupling, 22 x 15mm	nr	0.20	2.19	1.74	0.59	4.52
straight female connector, 22mm x 3/4in	nr	0.20	2.19	3.18	0.81	6.17
female reducing connector, 22mm x 1/2in	nr	0.20	2.19	5.05	1.09	8.32
female reducing connector, 22mm x 1in	nr	0.20	2.19	9.37	1.73	13.29
straight male connector, 22mm x 3/4in	nr	0.20	2.19	3.39	0.84	6.41
male reducing connector, 22mm x 1/2in	nr	0.20	2.19	8.30	1.57	12.06
male reducing connector, 22mm x 1in	nr	0.20	2.19	7.60	1.47	11.26
lead connector	nr	0.20	2.19	2.44	0.69	5.32
tank connector, 22mm x 1in	nr	0.20	2.19	6.94	1.37	10.50
reducer, 22 x 12mm	nr	0.20	2.19	2.37	0.68	5.24
female adaptor, 22 x 3/4in	nr	0.20	2.19	5.75	1.19	9.13
male adaptor, 22 x 3/4in	nr	0.20	2.19	4.93	1.07	8.19
adaptor, 22 x 3/4in	nr	0.20	2.19	1.77	0.59	4.55
reducing elbow, 22 x 15mm	nr	0.20	2.19	5.98	1.23	9.39
street elbow, 22mm	nr	0.20	2.19	1.61	0.57	4.37
male elbow, 22mm x 3/4in	nr	0.20	2.19	5.30	1.12	8.61
female elbow, 22mm x 3/4in	nr	0.20	2.19	4.66	1.03	7.88

	Unit	Hours	Hours £	Materials £	O & P £	Total £
reducing female elbow, 22mm x 1/2in	nr	0.20	2.19	5.36	1.13	8.68
backplate elbow, 22mm x 3/4in	nr	0.20	2.19	8.41	1.59	12.19
flanged bend, 22mm x 3/4in	nr	0.20	2.19	8.67	1.63	12.49
overflow bend, 22mm x 3/4in	nr	0.20	2.19	11.85	2.11	16.14
slow bend	nr	0.20	2.19	1.34	0.53	4.06
return bend	nr	0.20	2.19	10.07	1.84	14.10
obtuse elbow	nr	0.20	2.19	1.74	0.59	4.52
tee, reduced branch largest end 22mm	nr	0.24	2.63	5.07	1.15	8.85
tee, both ends reduced, largest end 22mm	nr	0.24	2.63	4.12	1.01	7.76
corner tee, 22mm	nr	0.24	2.63	10.69	2.00	15.31
sweep tee, 90 degrees, 22mm	nr	0.24	2.63	8.56	1.68	12.86
sweep tee, reduced branch, largest end 22mm	nr	0.24	2.63	7.14	1.46	11.23
offset tee, 22mm	nr	0.24	2.63	10.43	1.96	15.01
double sweep tee, 22mm	nr	0.24	2.63	10.17	1.92	14.71
cross equal tee, 22mm	nr	0.24	2.63	11.02	2.05	15.69
stop end, 22mm	nr	0.20	2.19	1.98	0.63	4.79
tap connector, 22mm x 1/2in	nr	0.20	2.19	7.48	1.45	11.12
bent tap connector, 22mm x 1/2in	nr	0.20	2.19	9.11	1.69	12.99
bent tap connector, 22mm x 3/4in	nr	0.20	2.19	4.19	0.96	7.33
bent union adaptor, 22mm x 1in	nr	0.20	2.19	6.02	1.23	9.44
bent male union adaptor, 22mm x 3/4in	nr	0.20	2.19	9.92	1.82	13.92
bent female union adaptor, 22mm x 3/4in	nr	0.20	2.19	9.92	1.82	13.92
made bend	nr	0.20	2.19	0.00	0.33	2.52

	Unit	Hours	Hours £	Materials £	O & P £	Total £
Stop valves						
Gunmetal stop valve with brass headwork, copper x copper						
15mm	nr	0.20	2.19	4.11	0.94	7.24
22mm	nr	0.22	2.41	7.67	1.51	11.59
28mm	nr	0.28	3.06	21.85	3.74	28.65
Gunmetal lockshield stop valve with brass headwork, copper x copper						
15mm	nr	0.20	2.19	10.09	1.84	14.12
22mm	nr	0.22	2.41	14.50	2.54	19.44
28mm	nr	0.28	3.06	26.59	4.45	34.10
Valves						
Radiator valve, chromium plated, compression inlet x taper male union outlet						
8mm x 1/2in	nr	0.30	3.28	8.20	1.72	13.20
10mm x 1/2in	nr	0.30	3.28	8.20	1.72	13.20
15mm x 1/2in	nr	0.30	3.28	8.20	1.72	13.20
Copper pipework, compression joints						
Copper pipe to BS 2871 Table X, compression joints and fittings to BS 864, clips at 1250mm maximum centres						
15mm diameter, to timber	m	0.20	2.19	1.60	0.57	4.36
15mm diameter, plugged and screwed	m	0.22	2.41	1.88	0.64	4.93
extra over for						
straight coupling, 15mm	nr	0.18	1.97	0.99	0.44	3.40
male coupling, extended thread, 15mm x 1/2in	nr	0.18	1.97	0.90	0.43	3.30

	Unit	Hours	Hours £	Materials £	O & P £	Total £
lead coupling, 15mm x 3/4in	nr	0.18	1.97	4.47	0.97	7.41
bent union radiator, 15mm x 1/2in	nr	0.18	1.97	5.44	1.11	8.52
air release elbow, 15mm	nr	0.18	1.97	6.16	1.22	9.35
slow bend, 15mm	nr	0.18	1.97	6.85	1.32	10.14
male elbow, 15mm x 1/2in	nr	0.18	1.97	1.90	0.58	4.45
male elbow, 15mm x 3/4in	nr	0.18	1.97	5.81	1.17	8.95
male elbow, BSP parallel thread, 15mm x 1/2in	nr	0.18	1.97	1.90	0.58	4.45
female elbow, 15mm x 1/2in	nr	0.18	1.97	2.92	0.73	5.62
female wall elbow, 15mm x 1/2in	nr	0.18	1.97	4.23	0.93	7.13
female wall elbow, 15mm x 3/4in	nr	0.18	1.97	9.57	1.73	13.27
female tee with backplate, 15 x 15 x 15mm	nr	0.18	1.97	8.02	1.50	11.49
offset tee, 15 x 15 x 15mm	nr	0.18	1.97	12.61	2.19	16.77
male adaptor, 15mm x 1/2in	nr	0.18	1.97	2.32	0.64	4.93
reducing set, 15 x 8mm	nr	0.18	1.97	1.57	0.53	4.07
reducing set, 15 x 10mm	nr	0.18	1.97	1.57	0.53	4.07
reducing set, 15 x 12mm	nr	0.18	1.97	1.57	0.53	4.07
22mm diameter, to timber	m	0.20	2.19	3.13	0.80	6.12
22mm diameter, plugged and screwed	m	0.22	2.41	3.39	0.87	6.67
extra over for						
straight coupling, 22mm	nr	0.18	1.97	1.70	0.55	4.22
male coupling, extended thread, 22mm x 3/4in	nr	0.18	1.97	1.42	0.51	3.90
female coupling, 22mm x 1in	nr	0.18	1.97	2.00	0.60	4.56
lead coupling, 22mm	nr	0.18	1.97	4.94	1.04	7.95
air release elbow, 22mm	nr	0.18	1.97	8.11	1.51	11.59
slow bend, 22mm	nr	0.18	1.97	11.00	1.95	14.91
male elbow, 22mm x 3/4in	nr	0.18	1.97	2.46	0.66	5.09
male elbow, 22mm x 1in	nr	0.18	1.97	4.81	1.02	7.80
male elbow, BSP parallel thread, 22mm x 3/4in	nr	0.18	1.97	2.46	0.66	5.09
male elbow, BSP parallel thread, 22mm x 1in	nr	0.18	1.97	4.80	1.02	7.78

	Unit	Hours	Hours £	Materials £	O & P £	Total £
female elbow, 22mm x 3/4in	nr	0.18	1.97	4.23	0.93	7.13
female elbow, 22mm x 1in	nr	0.18	1.97	6.57	1.28	9.82
bent cylinder connector, 22mm x 1in	nr	0.18	1.97	9.57	1.73	13.27
tee with reduced branch, 22 x 22 x 15mm	nr	0.22	2.41	6.09	1.27	9.77
tee, end reduced, 22 x 15 x 22mm	nr	0.22	2.41	7.07	1.42	10.90
tee, end and branch reduced, 22 x 15 x 15mm	nr	0.22	2.41	6.84	1.39	10.63
straight swivel connector, 22mm x 1/2in	nr	0.18	1.97	5.71	1.15	8.83
bent swivel connector, 22mm x 1/2in	nr	0.18	1.97	7.54	1.43	10.94
offset tee, 22 x 22 x 15mm	nr	0.22	2.41	18.57	3.15	24.12
tank coupling, extended thread 22mm	nr	0.18	1.97	7.76	1.46	11.19

Stop valves

Gunmetal stop valve with brass headwork, copper x copper

	Unit	Hours	Hours £	Materials £	O & P £	Total £
15mm	nr	0.19	2.08	5.57	1.15	8.80
22mm	nr	0.21	2.30	9.84	1.82	13.96
28mm	nr	0.27	2.95	25.63	4.29	32.87

Gunmetal lockshield stop valve with brass headwork, copper x copper

	Unit	Hours	Hours £	Materials £	O & P £	Total £
15mm	nr	0.19	2.08	12.16	2.14	16.37
22mm	nr	0.21	2.30	17.17	2.92	22.39
28mm	nr	0.27	2.95	33.02	5.40	41.37

Cold water storage tanks

Galvanised steel cistern with cover, reference

	Unit	Hours	Hours £	Materials £	O & P £	Total £
SC10, 18 litres	nr	0.60	6.56	39.11	6.85	52.53
SC15, 36 litres	nr	0.65	7.11	44.79	7.79	59.69
SC20, 54 litres	nr	0.65	7.11	51.08	8.73	66.92

	Unit	Hours	Hours £	Materials £	O & P £	Total £
SC25, 68 litres	nr	0.70	7.66	53.02	9.10	69.78
SC30, 86 litres	nr	0.75	8.21	60.11	10.25	78.56
SC40, 114 litres	nr	0.80	8.75	62.54	10.69	81.99
Plastic cistern with lid, reference cover, reference						
PC4, 18 litres	nr	0.60	6.56	9.24	2.37	18.17
PC15, 68 litres	nr	0.70	7.66	31.64	5.89	45.19
PC20, 91 litres	nr	0.75	8.21	33.35	6.23	47.79
PC25, 114 litres	nr	0.80	8.75	42.96	7.76	59.47
PC40, 182 litres	nr	0.90	9.85	71.80	12.25	93.89
PC50, 227 litres	nr	1.00	10.94	76.89	13.17	101.00
Cutting holes for connectors						
13mm	nr	0.10	1.09	0.00	0.16	1.26
18mm	nr	0.12	1.31	0.00	0.20	1.51
25mm	nr	0.15	1.64	0.00	0.25	1.89
32mm	nr	0.20	2.19	0.00	0.33	2.52

Hot water tanks

	Unit	Hours	Hours £	Materials £	O & P £	Total £
Galvanised steel tank with cover, reference						
T25/1, 36 litres	nr	0.65	7.11	127.36	20.17	154.64
T25/2, 36 litres	nr	0.65	7.11	129.34	20.47	156.92
T30/1, 86 litres	nr	0.75	8.21	136.44	21.70	166.34
T30/2, 36 litres	nr	0.75	8.21	138.12	21.95	168.27
T40, 114 litres	nr	0.80	8.75	167.29	26.41	202.45

Hot water copper cylinders

	Unit	Hours	Hours £	Materials £	O & P £	Total £
Direct cylinders, insulated, reference						
ref 5, 98 litres	nr	0.50	5.47	75.24	12.11	92.82
ref 7, 120 litres	nr	0.55	6.02	84.01	13.50	103.53
ref 8, 148 litres	nr	0.65	7.11	91.66	14.82	113.59
ref 9, 166 litres	nr	0.85	9.30	107.58	17.53	134.41

	Unit	Hours	Hours £	Materials £	O & P £	Total £
Indirect cylinders, insulated, reference						
ref 2, 96 litres	nr	0.50	5.47	91.41	14.53	111.41
ref 3, 114 litres	nr	0.55	6.02	97.30	15.50	118.81
ref 7, 117 litres	nr	0.55	6.02	96.28	15.34	117.64
ref 8, 140 litres	nr	0.65	7.11	108.71	17.37	133.19
ref 9, 162 litres	nr	0.70	7.66	142.19	22.48	172.33
900 x 450mm, capacity 85 litres hot, 25 litres cold, reference						
F1, direct	nr	3.00	32.82	125.39	23.73	181.94
F3, indirect coil	nr	3.00	32.82	157.24	28.51	218.57
F6, primatic indirect	nr	3.00	32.82	164.89	29.66	227.37
F8, primatic indirect	nr	3.00	32.82	174.29	31.07	238.18
1075 x 450mm, capacity 115 litres hot, 25 litres cold, reference						
F1, direct	nr	3.20	35.01	137.92	25.94	198.87
F1, indirect coil	nr	3.20	35.01	171.03	30.91	236.94
F6, primatic indirect	nr	3.20	35.01	179.94	32.24	247.19
F8, primatic indirect	nr	3.20	35.01	189.72	33.71	258.44
1200 x 450mm, capacity 115 litres hot, 45 litres cold, reference						
F1, direct	nr	3.40	37.20	146.83	27.60	211.63
F1, indirect coil	nr	3.40	37.20	179.81	32.55	249.56
F6, primatic indirect	nr	3.40	37.20	188.33	33.83	259.35
F8, primatic indirect	nr	3.40	37.20	198.87	35.41	271.48
1300 x 450mm, capacity 130 litres hot, 45 litres cold, reference						
F1, direct	nr	3.60	39.38	171.14	31.58	242.10
F1, indirect coil	nr	3.60	39.38	201.00	36.06	276.44
F6, primatic indirect	nr	3.60	39.38	221.06	39.07	299.51
F8, primatic indirect	nr	3.60	39.38	231.97	40.70	312.06

	Unit	Hours	Hours £	Materials £	O & P £	Total £
1200 x 500mm, capacity 150 litres hot, 45 litres cold, reference						
F1, direct	nr	3.80	41.57	201.54	36.47	279.58
F1, indirect coil	nr	3.80	41.57	238.62	42.03	322.22
F6, primatic indirect	nr	3.80	41.57	253.41	44.25	339.23
F8, primatic indirect	nr	3.80	41.57	263.45	45.75	350.78
1400 x 500mm, capacity 115 litres hot, 115 litres cold, reference						
F1, direct	nr	4.00	43.76	222.57	39.95	306.28
F1, indirect coil	nr	4.00	43.76	207.64	37.71	289.11
F6, primatic indirect	nr	4.00	43.76	266.20	46.49	356.45
F8, primatic indirect	nr	4.00	43.76	275.60	47.90	367.26

Insulation

	Unit	Hours	Hours £	Materials £	O & P £	Total £
Preformed pipe lagging, fire-retardant foam 13mm thick for pipe size						
15mm	m	0.08	0.88	2.18	0.46	3.51
22mm	m	0.09	0.98	2.71	0.55	4.25
28mm	m	0.10	1.09	3.17	0.64	4.90
Preformed pipe lagging, fire-retardant foam 25mm thick for pipe size						
15mm	m	0.10	1.09	3.64	0.71	5.44
22mm	m	0.11	1.20	4.45	0.85	6.50
28mm	m	0.12	1.31	5.65	1.04	8.01
50mm glass fibre filled insulating jacket, fixing bands to tank size						
445 x 305 x 300mm	nr	0.35	3.83	2.62	0.97	7.42
495 x 368 x 362mm	nr	0.40	4.38	3.12	1.12	8.62
630 x 450 x 420mm	nr	0.45	4.92	4.30	1.38	10.61
665 x 490 x 515mm	nr	0.50	5.47	5.30	1.62	12.39
700 x 540 x 535mm	nr	0.60	6.56	6.82	2.01	15.39
995 x 605 x 595mm	nr	0.70	7.66	8.93	2.49	19.08

	Unit	Hours	Hours £	Materials £	O & P £	Total £
60mm glass fibre filled insulating jacket, fixing bands to cylinder 450mm diameter, height						
750mm	nr	0.30	3.28	6.02	1.40	10.70
900mm	nr	0.35	3.83	6.50	1.55	11.88
1050mm	nr	0.40	4.38	7.40	1.77	13.54

Boilers

Gas-fired wall-mounted boilers
for domestic central heating
and indirect hot water

Balanced flue, output

	Unit	Hours	Hours £	Materials £	O & P £	Total £
20,000-30,000 BTU	nr	4.80	52.51	480.57	79.96	613.04
31,000-40,000 BTU	nr	4.80	52.51	549.22	90.26	691.99
41,000-50,000 BTU	nr	4.80	52.51	621.23	101.06	774.80
51,000-60,000 BTU	nr	4.80	52.51	775.42	124.19	952.12

Powered flue, output

	Unit	Hours	Hours £	Materials £	O & P £	Total £
20,000-30,000 BTU	nr	4.20	45.95	569.56	92.33	707.83
31,000-40,000 BTU	nr	4.20	45.95	639.79	102.86	788.60
41,000-50,000 BTU	nr	4.20	45.95	695.51	111.22	852.68
51,000-60,000 BTU	nr	4.20	45.95	754.59	120.08	920.62
61,000-80,000 BTU	nr	4.20	45.95	1051.09	164.56	1261.59

Gas-fired floor-standing boilers
for domestic central heating
and indirect hot water

Open flue, output

	Unit	Hours	Hours £	Materials £	O & P £	Total £
31,000-40,000 BTU	nr	4.80	52.51	517.20	85.46	655.17
41,000-50,000 BTU	nr	4.80	52.51	543.30	89.37	685.18
51,000-60,000 BTU	nr	4.80	52.51	583.64	95.42	731.57
61,000-70,000 BTU	nr	4.80	52.51	683.27	110.37	846.15
71,000-80,000 BTU	nr	4.80	52.51	747.33	119.98	919.82

	Unit	Hours	Hours £	Materials £	O & P £	Total £

Oil-fired boiler for domestic central heating and indirect hot water

Balanced flue, output

	Unit	Hours	Hours £	Materials £	O & P £	Total £
52,000 BTU	nr	4.80	52.51	1148.00	180.08	1380.59
70,000 BTU	nr	4.80	52.51	1178.75	184.69	1415.95
90,000 BTU	nr	4.80	52.51	1322.25	206.21	1580.98
120,000 BTU	nr	4.80	52.51	1501.63	233.12	1787.26
140,000 BTU	nr	4.80	52.51	1614.38	250.03	1916.93

Conventional flue, output

	Unit	Hours	Hours £	Materials £	O & P £	Total £
52,000 BTU	nr	4.20	45.95	986.55	154.87	1187.37
70,000 BTU	nr	4.20	45.95	1162.50	181.27	1389.72
90,000 BTU	nr	4.20	45.95	1247.36	194.00	1487.30
120,000 BTU	nr	4.20	45.95	1402.78	217.31	1666.04
140,000 BTU	nr	4.20	45.95	1521.64	235.14	1802.73

Fibre-cement light-quality flue pipe and fittings, joints in running lengths, pipe supports fixed to brickwork, pipe size

	Unit	Hours	Hours £	Materials £	O & P £	Total £
75mm	m	0.30	3.28	14.11	2.61	20.00
extra over for						
adjustable bend	nr	0.35	3.83	10.18	2.10	16.11
equal tee	nr	0.45	4.92	22.48	4.11	31.51
cone cap	nr	0.35	3.83	22.43	3.94	30.20
cast aluminium GC1 terminal	nr	0.35	3.83	26.18	4.50	34.51
blank cap	nr	0.35	3.83	2.32	0.92	7.07
100mm	m	0.35	3.83	14.83	2.80	21.46
extra over for						
adjustable bend	nr	0.40	4.38	11.36	2.36	18.10
equal tee	nr	0.45	4.92	22.63	4.13	31.69
cone cap	nr	0.40	4.38	23.24	4.14	31.76
cast aluminium GC1 terminal	nr	0.40	4.38	25.97	4.55	34.90
blank cap	nr	0.40	4.38	2.39	1.01	7.78

	Unit	Hours	Hours £	Materials £	O & P £	Total £
125mm	m	0.40	4.38	18.23	3.39	26.00
extra over for						
adjustable bend	nr	0.45	4.92	13.44	2.75	21.12
equal tee	nr	0.50	5.47	24.40	4.48	34.35
cone cap	nr	0.45	4.92	24.40	4.40	33.72
cast aluminium GC1 terminal	nr	0.45	4.92	28.60	5.03	38.55
blank cap	nr	0.45	4.92	2.61	1.13	8.66
150mm	m	0.45	4.92	19.56	3.67	28.16
extra over for						
adjustable bend	nr	0.50	5.47	16.76	3.33	25.56
equal tee	nr	0.55	6.02	25.20	4.68	35.90
cone cap	nr	0.50	5.47	28.74	5.13	39.34
blank cap	nr	0.50	5.47	2.79	1.24	9.50

Oil storage tanks

Standard domestic pattern oil
storage tank with 50mm hinged
filler, draw-off and sludge
sockets, 14g galvanised steel
plate

	Unit	Hours	Hours £	Materials £	O & P £	Total £
4 x 3 x 3 ft, 250 gallon capacity	nr	1.75	19.15	151.48	25.59	196.22
5 x 2 x 4 ft, 250 gallon capacity	nr	1.75	19.15	151.48	25.59	196.22
4 x 4 x 4 ft, 500 gallon capacity	nr	2.75	30.09	190.48	33.08	253.65
6 x 4 x 4 ft, 600 gallon capacity	nr	3.00	32.82	238.48	40.70	312.00

Standard domestic pattern oil
storage tank with 50mm hinged
filler, draw-off and sludge
sockets, 12g galvanised steel
plate

	Unit	Hours	Hours £	Materials £	O & P £	Total £
4 x 3 x 3 ft, 250 gallon capacity	nr	1.75	19.15	173.98	28.97	222.09
5 x 2 x 4 ft, 300 gallon capacity	nr	2.25	24.62	173.98	29.79	228.38
4 x 3 x 4 ft, 300 gallon capacity	nr	2.25	24.62	194.98	32.94	252.53
6 x 4 x 4 ft, 600 gallon capacity	nr	3.00	32.82	272.97	45.87	351.66
8 x 4 x 4 ft, 800 gallon capacity	nr	3.75	41.03	368.96	61.50	471.48

	Unit	Hours	Hours £	Materials £	O & P £	Total £
Standard domestic pattern oil storage tank with 50mm hinged filler, draw-off and sludge sockets, 10g galvanised steel plate						
4 x 4 x 4 ft, 400 gallon capacity	nr	2.50	27.35	238.48	39.87	305.70
5 x 4 x 4 ft, 500 gallon capacity	nr	2.75	30.09	256.46	42.98	329.53
6 x 4 x 4 ft, 600 gallon capacity	nr	3.00	32.82	314.96	52.17	399.95
7 x 4 x 4 ft, 700 gallon capacity	nr	3.50	38.29	361.66	59.99	459.94
8 x 4 x 4 ft, 800 gallon capacity	nr	3.75	41.03	412.45	68.02	521.50
8 x 5 x 4 ft, 1000 gallon capacity	nr	4.00	43.76	505.44	82.38	631.58
8 x 6 x 4 ft, 1200 gallon capacity	nr	4.50	49.23	583.43	94.90	727.56
8 x 5 x 5 ft, 1250 gallon capacity	nr	4.50	49.23	586.43	95.35	731.01
8 x 6 x 6 ft, 1300 gallon capacity	nr	4.75	51.97	605.93	98.68	756.58
Extra for tank stand up to 600mm high						
100 gallon tank	nr	1.00	10.94	92.76	15.56	119.26
250 gallon tank	nr	1.50	16.41	94.36	16.62	127.39
400 gallon tank	nr	2.00	21.88	105.55	19.11	146.54
500 gallon tank	nr	2.00	21.88	110.35	19.83	152.06
600 gallon tank	nr	2.00	21.88	116.75	20.79	159.42
700 gallon tank	nr	2.50	27.35	123.14	22.57	173.06
800 gallon tank	nr	2.50	27.35	143.93	25.69	196.97
1000 gallon tank	nr	3.00	32.82	156.73	28.43	217.98
1250 gallon tank	nr	3.00	32.82	169.52	30.35	232.69
Dipstick, flat mild steel section						
3 ft depth	nr	0.50	5.47	17.53	3.45	26.45
4 ft depth	nr	0.50	5.47	19.80	3.79	29.06
5 ft depth	nr	0.50	5.47	21.03	3.98	30.48
6 ft depth	nr	0.75	8.21	23.09	4.69	35.99
8 ft depth	nr	0.75	8.21	26.30	5.18	39.68

	Unit	Hours	Hours £	Materials £	O & P £	Total £

Radiators

Single panel radiator on
brackets plugged and screwed
to brickwork

450mm high, length

500mm	nr	1.00	10.94	18.98	4.49	34.41
1000mm	nr	1.15	12.58	37.98	7.58	58.15
1600mm	nr	1.25	13.68	60.75	11.16	85.59
2000mm	nr	1.40	15.32	75.09	13.56	103.97
2400mm	nr	1.50	16.41	91.12	16.13	123.66

600mm high, length

500mm	nr	1.10	12.03	25.44	5.62	43.10
1000mm	nr	1.25	13.68	50.85	9.68	74.20
1600mm	nr	1.35	14.77	81.37	14.42	110.56
2000mm	nr	1.50	16.41	101.72	17.72	135.85
2400mm	nr	1.60	17.50	122.07	20.94	160.51

700mm high, length

500mm	nr	1.20	13.13	29.76	6.43	49.32
1000mm	nr	1.35	14.77	59.50	11.14	85.41
1600mm	nr	1.45	15.86	95.18	16.66	127.70
2000mm	nr	1.60	17.50	118.97	20.47	156.95
2400mm	nr	1.70	18.60	142.80	24.21	185.61

Single panel single covector
radiator fixed to concealed
brackets plugged and
and screwed to brickwork

300mm high, length

500mm	nr	0.90	9.85	19.75	4.44	34.04
1000mm	nr	1.05	11.49	39.49	7.65	58.62
1500mm	nr	1.15	12.58	59.24	10.77	82.59
2000mm	nr	1.30	14.22	78.98	13.98	107.18
2500mm	nr	1.40	15.32	98.73	17.11	131.15

	Unit	Hours	Hours £	Materials £	O & P £	Total £
450mm high, length						
500mm	nr	1.00	10.94	18.43	4.41	33.78
1000mm	nr	1.15	12.58	33.19	6.87	52.64
1400mm	nr	1.25	13.68	51.62	9.79	75.09
2000mm	nr	1.40	15.32	113.53	19.33	148.17
2600mm	nr	1.50	16.41	147.56	24.60	188.57
600mm high, length						
500mm	nr	1.10	12.03	23.80	5.38	41.21
1000mm	nr	1.25	13.68	47.60	9.19	70.47
1400mm	nr	1.35	14.77	66.64	12.21	93.62
2000mm	nr	1.50	16.41	146.58	24.45	187.44
2600mm	nr	1.60	17.50	190.54	31.21	239.25

Double panel single convector radiator fixed to concealed brackets plugged and screwed to brickwork

	Unit	Hours	Hours £	Materials £	O & P £	Total £
300mm high, length						
500mm	nr	0.90	9.85	31.37	6.18	47.40
1000mm	nr	1.05	11.49	62.72	11.13	85.34
1500mm	nr	1.15	12.58	94.09	16.00	122.67
2000mm	nr	1.30	14.22	125.43	20.95	160.60
2500mm	nr	1.40	15.32	156.79	25.82	197.92
450mm high, length						
500mm	nr	1.00	10.94	28.64	5.94	45.52
1000mm	nr	1.15	12.58	57.29	10.48	80.35
1400mm	nr	1.25	13.68	91.81	15.82	121.31
2000mm	nr	1.40	15.32	176.40	28.76	220.47
2600mm	nr	1.50	16.41	229.33	36.86	282.60
600mm high, length						
500mm	nr	1.10	12.03	36.10	7.22	55.35
1000mm	nr	1.25	13.68	72.19	12.88	98.74
1400mm	nr	1.35	14.77	115.68	19.57	150.02
2000mm	nr	1.50	16.41	222.28	35.80	274.49
2600mm	nr	1.60	17.50	288.96	45.97	352.43

	Unit	Hours	Hours £	Materials £	O & P £	Total £
Double panel double convector radiator fixed to concealed brackets plugged and screwed to brickwork						
300mm high, length						
500mm	nr	0.90	9.85	39.11	7.34	56.30
1000mm	nr	1.05	11.49	78.25	13.46	103.20
1500mm	nr	1.15	12.58	117.37	19.49	149.44
2000mm	nr	1.30	14.22	156.49	25.61	196.32
2500mm	nr	1.40	15.32	195.61	31.64	242.56
450mm high, length						
500mm	nr	1.00	10.94	34.77	6.86	52.57
1000mm	nr	1.15	12.58	69.54	12.32	94.44
1400mm	nr	1.25	13.68	111.43	18.77	143.87
2000mm	nr	1.40	15.32	214.10	34.41	263.83
2600mm	nr	1.50	16.41	278.33	44.21	338.95
600mm high, length						
500mm	nr	1.10	12.03	43.78	8.37	64.19
1000mm	nr	1.25	13.68	87.57	15.19	116.43
1400mm	nr	1.35	14.77	140.32	23.26	178.35
2000mm	nr	1.50	16.41	269.61	42.90	328.92
2600mm	nr	1.60	17.50	350.50	55.20	423.20

	Unit	Hours	Hours £	Materials £	O & P £	Total £

GLAZING

Clear float glass

In wood with putty and sprigs

under 0.15m2, thickness

	Unit	Hours	Hours £	Materials £	O & P £	Total £
3mm	m2	0.90	9.10	23.48	4.89	37.47
4mm	m2	0.90	9.10	25.08	5.13	39.31
5mm	m2	0.90	9.10	28.87	5.70	43.66
6mm	m2	1.00	10.11	30.55	6.10	46.76
10mm	m2	1.05	10.62	52.37	9.45	72.43

over 0.15m2, thickness

	Unit	Hours	Hours £	Materials £	O & P £	Total £
3mm	m2	0.60	6.07	23.48	4.43	33.98
4mm	m2	0.60	6.07	25.08	4.67	35.82
5mm	m2	0.60	6.07	28.87	5.24	40.18
6mm	m2	0.65	6.57	30.55	5.57	42.69
10mm	m2	0.70	7.08	52.37	8.92	68.36

In wood with pinned beads

under 0.15m2, thickness

	Unit	Hours	Hours £	Materials £	O & P £	Total £
3mm	m2	1.10	11.12	23.48	5.19	39.79
4mm	m2	1.10	11.12	25.08	5.43	41.63
5mm	m2	1.10	11.12	28.87	6.00	45.99
6mm	m2	1.20	12.13	30.55	6.40	49.08
10mm	m2	1.25	12.64	52.37	9.75	74.76

over 0.15m2, thickness

	Unit	Hours	Hours £	Materials £	O & P £	Total £
3mm	m2	0.80	8.09	23.48	4.74	36.30
4mm	m2	0.80	8.09	25.08	4.98	38.14
5mm	m2	0.80	8.09	28.87	5.54	42.50
6mm	m2	0.90	9.10	30.55	5.95	45.60
10mm	m2	0.95	9.60	52.37	9.30	71.27

	Unit	Hours	Hours £	Materials £	O & P £	Total £
In wood with screwed beads						
under 0.15m2, thickness						
3mm	m2	1.30	13.14	23.48	5.49	42.12
4mm	m2	1.30	13.14	25.08	5.73	43.96
5mm	m2	1.30	13.14	28.87	6.30	48.31
6mm	m2	1.40	14.15	30.55	6.71	51.41
10mm	m2	1.45	14.66	52.37	10.05	77.08
over 0.15m2, thickness						
3mm	m2	1.00	10.11	23.48	5.04	38.63
4mm	m2	1.00	10.11	25.08	5.28	40.47
5mm	m2	1.00	10.11	28.87	5.85	44.83
6mm	m2	1.05	10.62	30.55	6.17	47.34
10mm	m2	1.10	11.12	52.37	9.52	73.01
In metal with putty						
under 0.15m2, thickness						
3mm	m2	0.90	9.10	23.48	4.89	37.47
4mm	m2	0.90	9.10	25.08	5.13	39.31
5mm	m2	0.90	9.10	28.87	5.70	43.66
6mm	m2	1.00	10.11	30.55	6.10	46.76
10mm	m2	1.05	10.62	52.37	9.45	72.43
over 0.15m2, thickness						
3mm	m2	0.60	6.07	23.48	4.43	33.98
4mm	m2	0.60	6.07	25.08	4.67	35.82
5mm	m2	0.60	6.07	28.87	5.24	40.18
6mm	m2	0.65	6.57	30.55	5.57	42.69
10mm	m2	0.70	7.08	52.37	8.92	68.36
In metal with clips and putty						
under 0.15m2, thickness						
3mm	m2	0.95	9.60	23.48	4.96	38.05
4mm	m2	0.95	9.60	25.08	5.20	39.89
5mm	m2	0.95	9.60	28.87	5.77	44.25
6mm	m2	1.05	10.62	30.55	6.17	47.34
10mm	m2	1.10	11.12	52.37	9.52	73.01

	Unit	Hours	Hours £	Materials £	O & P £	Total £
over 0.15m2, thickness						
3mm	m2	0.65	6.57	23.48	4.51	34.56
4mm	m2	0.65	6.57	25.08	4.75	36.40
5mm	m2	0.65	6.57	28.87	5.32	40.76
6mm	m2	0.70	7.08	30.55	5.64	43.27
10mm	m2	0.75	7.58	52.37	8.99	68.95

In metal with screwed metal beads

	Unit	Hours	Hours £	Materials £	O & P £	Total £
under 0.15m2, thickness						
3mm	m2	1.30	13.14	23.48	5.49	42.12
4mm	m2	1.30	13.14	25.08	5.73	43.96
5mm	m2	1.30	13.14	28.87	6.30	48.31
6mm	m2	1.40	14.15	30.55	6.71	51.41
10mm	m2	1.45	14.66	52.37	10.05	77.08
over 0.15m2, thickness						
3mm	m2	1.00	10.11	23.48	5.04	38.63
4mm	m2	1.00	10.11	25.08	5.28	40.47
5mm	m2	1.00	10.11	28.87	5.85	44.83
6mm	m2	1.05	10.62	30.55	6.17	47.34
10mm	m2	1.10	11.12	52.37	9.52	73.01

White patterned glass

In wood with putty and sprigs

	Unit	Hours	Hours £	Materials £	O & P £	Total £
under 0.15m2, thickness						
4mm	m2	0.90	9.10	27.65	5.51	42.26
6mm	m2	1.00	10.11	32.44	6.38	48.93
over 0.15m2, thickness						
4mm	m2	0.60	6.07	27.65	5.06	38.77
6mm	m2	0.65	6.57	32.44	5.85	44.86

In wood with pinned beads

	Unit	Hours	Hours £	Materials £	O & P £	Total £
under 0.15m2, thickness						
4mm	m2	1.10	11.12	27.65	5.82	44.59
6mm	m2	1.20	12.13	32.44	6.69	51.26

	Unit	Hours	Hours £	Materials £	O & P £	Total £
over 0.15m2, thickness						
4mm	m2	0.80	8.09	27.65	5.36	41.10
6mm	m2	0.85	8.59	32.44	6.16	47.19
In wood with screwed beads						
under 0.15m2, thickness						
4mm	m2	1.30	13.14	27.65	6.12	46.91
6mm	m2	1.40	14.15	32.44	6.99	53.58
over 0.15m2, thickness						
4mm	m2	1.00	10.11	27.65	5.66	43.42
6mm	m2	1.50	15.17	32.44	7.14	54.75
In metal with putty						
under 0.15m2, thickness						
4mm	m2	0.90	9.10	27.65	5.51	42.26
6mm	m2	1.00	10.11	32.44	6.38	48.93
over 0.15m2, thickness						
4mm	m2	0.60	6.07	27.65	5.06	38.77
6mm	m2	0.65	6.57	32.44	5.85	44.86
In metal with clips and putty						
under 0.15m2, thickness						
4mm	m2	0.95	9.60	27.65	5.59	42.84
6mm	m2	1.05	10.62	32.44	6.46	49.51
over 0.15m2, thickness						
4mm	m2	0.65	6.57	27.65	5.13	39.35
6mm	m2	0.70	7.08	32.44	5.93	45.44
In metal with screwed metal beads						
under 0.15m2, thickness						
4mm	m2	1.30	13.14	27.65	6.12	46.91
6mm	m2	1.40	14.15	32.44	6.99	53.58

	Unit	Hours	Hours £	Materials £	O & P £	Total £
over 0.15m2, thickness						
4mm	m2	1.00	10.11	27.65	5.66	43.42
6mm	m2	1.05	10.62	32.44	6.46	49.51

Georgian wired cast glass

In wood with putty and sprigs

	Unit	Hours	Hours £	Materials £	O & P £	Total £
under 0.15m2, thickness						
7mm	m2	0.90	9.10	28.78	5.68	43.56
over 0.15m2, thickness						
7mm	m2	0.65	6.57	28.78	5.30	40.65

In wood with pinned beads

	Unit	Hours	Hours £	Materials £	O & P £	Total £
under 0.15m2, thickness						
7mm	m2	1.20	12.13	28.78	6.14	47.05
over 0.15m2, thickness						
7mm	m2	0.90	9.10	28.78	5.68	43.56

In wood with screwed beads

	Unit	Hours	Hours £	Materials £	O & P £	Total £
under 0.15m2, thickness						
7mm	m2	1.40	14.15	28.78	6.44	49.37
over 0.15m2, thickness						
7mm	m2	1.05	10.62	28.78	5.91	45.30

In metal with putty and clips

	Unit	Hours	Hours £	Materials £	O & P £	Total £
under 0.15m2, thickness						
7mm	m2	1.05	10.62	28.78	5.91	45.30
over 0.15m2, thickness						
7mm	m2	0.70	7.08	28.78	5.38	41.24

	Unit	Hours	Hours £	Materials £	O & P £	Total £
In metal with screwed metal beads						
under 0.15m2, thickness 7mm	m2	1.40	14.15	28.78	6.44	49.37
over 0.15m2, thickness 7mm	m2	1.05	10.62	28.78	5.91	45.30
Georgian wired polished glass						
In wood with putty and sprigs						
under 0.15m2, thickness 6mm	m2	0.90	9.10	67.89	11.55	88.54
over 0.15m2, thickness 6mm	m2	0.65	6.57	67.89	11.17	85.63
In wood with pinned beads						
under 0.15m2, thickness 6mm	m2	1.20	12.13	67.89	12.00	92.03
over 0.15m2, thickness 6mm	m2	0.90	9.10	67.89	11.55	88.54
In wood with screwed beads						
under 0.15m2, thickness 6mm	m2	1.40	14.15	67.89	12.31	94.35
over 0.15m2, thickness 6mm	m2	1.05	10.62	67.89	11.78	90.28

	Unit	Hours	Hours £	Materials £	O & P £	Total £
In metal with putty and clips						
under 0.15m2, thickness						
7mm	m2	1.05	10.62	67.89	11.78	90.28
over 0.15m2, thickness						
7mm	m2	0.70	7.08	67.89	11.25	86.21
In metal with screwed metal beads						
under 0.15m2, thickness						
7mm	m2	1.40	14.15	67.89	12.31	94.35
over 0.15m2, thickness						
7mm	m2	1.05	10.62	67.89	11.78	90.28
Clear laminated safety glass						
In wood with pinned beads						
under 0.15m2, thickness						
4mm	m2	1.10	11.12	48.55	8.95	68.62
4.4mm	m2	1.10	11.12	54.27	9.81	75.20
6.4mm	m2	1.20	12.13	58.67	10.62	81.42
over 0.15m2, thickness						
4mm	m2	0.80	8.09	48.55	8.50	65.13
4.4mm	m2	0.80	8.09	54.27	9.35	71.71
6.4mm	m2	0.90	9.10	58.67	10.17	77.93
In wood with screwed beads						
under 0.15m2, thickness						
4mm	m2	1.30	13.14	48.55	9.25	70.95
4.4mm	m2	1.30	13.14	54.27	10.11	77.52
6.4mm	m2	1.40	14.15	58.67	10.92	83.75

	Unit	Hours	Hours £	Materials £	O & P £	Total £
over 0.15m2, thickness						
4mm	m2	1.00	10.11	48.55	8.80	67.46
4.4mm	m2	1.00	10.11	54.27	9.66	74.04
6.4mm	m2	1.05	10.62	58.67	10.39	79.68
In metal with screwed beads						
under 0.15m2, thickness						
4mm	m2	1.30	13.14	48.55	9.25	70.95
4.4mm	m2	1.30	13.14	54.27	10.11	77.52
6.4mm	m2	1.40	14.15	58.67	10.92	83.75
over 0.15m2, thickness						
4mm	m2	1.00	10.11	48.55	8.80	67.46
4.4mm	m2	1.00	10.11	54.27	9.66	74.04
6.4mm	m2	1.05	10.62	58.67	10.39	79.68

	Unit	Hours	Hours £	Materials £	O & P £	Total £

PAINTING AND WALLPAPERING

On internal surfaces before fixing

One coat of primer on wood
surfaces

	Unit	Hours	Hours £	Materials £	O & P £	Total £
exceeding 300mm girth	m2	0.16	1.62	0.48	0.31	2.41
not exceeding 150mm girth	m	0.04	0.40	0.08	0.07	0.56
150-300mm girth	m	0.06	0.61	0.16	0.11	0.88
isolated areas not exceeding						
0.5m2	nr	0.10	1.01	0.24	0.19	1.44

One coat of primer on metal
surfaces

	Unit	Hours	Hours £	Materials £	O & P £	Total £
exceeding 300mm girth	m2	0.16	1.62	0.62	0.34	2.57
not exceeding 150mm girth	m	0.04	0.40	0.10	0.08	0.58
150-300mm girth	m	0.06	0.61	0.20	0.12	0.93
isolated areas not exceeding						
0.5m2	nr	0.10	1.01	0.31	0.20	1.52

One coat of wood preservative
on wood surfaces

	Unit	Hours	Hours £	Materials £	O & P £	Total £
exceeding 300mm girth	m2	0.14	1.42	0.32	0.26	2.00
not exceeding 150mm girth	m	0.03	0.30	0.08	0.06	0.44
150-300mm girth	m	0.05	0.51	0.12	0.09	0.72
isolated areas not exceeding						
0.5m2	nr	0.08	0.81	0.16	0.15	1.11

On internal walls and ceilings

One coat of emulsion paint on

	Unit	Hours	Hours £	Materials £	O & P £	Total £
brickwork or blockwork walls	m2	0.20	2.02	0.60	0.39	3.02
brickwork or blockwork walls						
in staircase areas	m2	0.26	2.63	0.60	0.48	3.71
plastered walls	m2	0.12	1.21	0.40	0.24	1.86
plastered walls in staircase areas	m2	0.16	1.62	0.40	0.30	2.32

	Unit	Hours	Hours £	Materials £	O & P £	Total £
plastered ceilings	m2	0.20	2.02	0.40	0.36	2.79
plastered ceilings in staircase areas	m2	0.24	2.43	0.40	0.42	3.25
embossed papered walls	m2	0.16	1.62	0.50	0.32	2.44
embossed papered walls in staircase areas	m2	0.20	2.02	0.50	0.38	2.90

Two coats of emulsion paint on

	Unit	Hours	Hours £	Materials £	O & P £	Total £
brickwork or blockwork walls	m2	0.28	2.83	1.20	0.60	4.64
brickwork or blockwork walls in staircase areas	m2	0.32	3.24	1.20	0.67	5.10
plastered walls	m2	0.16	1.62	0.80	0.36	2.78
plastered walls in staircase areas	m2	0.20	2.02	0.80	0.42	3.25
plastered ceilings	m2	0.24	2.43	0.80	0.48	3.71
plastered ceilings in staircase areas	m2	0.28	2.83	0.80	0.54	4.18
embossed papered walls	m2	0.24	2.43	1.00	0.51	3.94
embossed papered walls in staircase areas	m2	0.26	2.63	1.00	0.54	4.17

One mist coat and two coats of emulsion paint on

	Unit	Hours	Hours £	Materials £	O & P £	Total £
brickwork or blockwork walls	m2	0.36	3.64	1.50	0.77	5.91
brickwork or blockwork walls in staircase areas	m2	0.40	4.04	1.50	0.83	6.38
plastered walls	m2	0.24	2.43	1.00	0.51	3.94
plastered walls in staircase areas	m2	0.28	2.83	1.00	0.57	4.41
plastered ceilings	m2	0.32	3.24	1.00	0.64	4.87
plastered ceilings in staircase areas	m2	0.36	3.64	1.00	0.70	5.34
embossed papered walls	m2	0.30	3.03	1.25	0.64	4.93
embossed papered walls in staircase areas	m2	0.34	3.44	1.25	0.70	5.39

	Unit	Hours	Hours £	Materials £	O & P £	Total £
One coat sealer primer, one undercoat and one coat gloss on						
plastered walls	m2	0.42	4.25	1.20	0.82	6.26
plastered walls in staircase areas	m2	0.46	4.65	1.20	0.88	6.73
plastered ceilings	m2	0.48	4.85	1.20	0.91	6.96
plastered ceilings in staircase areas	m2	0.52	5.26	1.20	0.97	7.43

On internal woodwork

One coat primer on wood
general surfaces

	Unit	Hours	Hours £	Materials £	O & P £	Total £
exceeding 300mm girth	m2	0.18	1.82	0.50	0.35	2.67
not exceeding 150mm girth	m	0.04	0.40	0.08	0.07	0.56
150-300mm girth	m	0.06	0.61	0.16	0.11	0.88
isolated areas not exceeding 0.5m2	nr	0.10	1.01	0.24	0.19	1.44

One coat primer on wood
windows, glazed doors and the
like

	Unit	Hours	Hours £	Materials £	O & P £	Total £
panes area not exceeding 0.1m2	m2	0.35	3.54	0.18	0.56	4.28
panes area 0.1 to 0.5m2	m2	0.30	3.03	0.16	0.48	3.67
panes area 0.5 to 1m2	m2	0.25	2.53	0.14	0.40	3.07
panes area exceeding 1m2	m2	0.22	2.22	0.12	0.35	2.70

One coat primer on wood frames,
linings, skirtings and the like

	Unit	Hours	Hours £	Materials £	O & P £	Total £
exceeding 300mm girth	m2	0.20	2.02	0.46	0.37	2.85
not exceeding 150mm girth	m	0.05	0.51	0.06	0.08	0.65
150-300mm girth	m	0.08	0.81	0.14	0.14	1.09
isolated areas not exceeding 0.5m2	nr	0.11	1.11	0.22	0.20	1.53

	Unit	Hours	Hours £	Materials £	O & P £	Total £
One coat undercoat on primed wood general surfaces						
exceeding 300mm girth	m2	0.18	1.82	0.50	0.35	2.67
not exceeding 150mm girth	m	0.04	0.40	0.08	0.07	0.56
150-300mm girth	m	0.06	0.61	0.16	0.11	0.88
isolated areas not exceeding 0.5m2	nr	0.10	1.01	0.24	0.19	1.44
One coat undercoat on primed wood windows, glazed doors and the like						
panes area not exceeding 0.1m2	m2	0.35	3.54	0.20	0.56	4.30
panes area 0.1 to 0.5m2	m2	0.30	3.03	0.18	0.48	3.69
panes area 0.5 to 1m2	m2	0.25	2.53	0.16	0.40	3.09
panes area exceeding 1m2	m2	0.22	2.22	0.14	0.35	2.72
One coat undercoat on primed wood frames, linings, skirtings and the like						
exceeding 300mm girth	m2	0.20	2.02	0.50	0.38	2.90
not exceeding 150mm girth	m	0.05	0.51	0.08	0.09	0.67
150-300mm girth	m	0.08	0.81	0.16	0.15	1.11
isolated areas not exceeding 0.5m2	nr	0.11	1.11	0.24	0.20	1.55
Two coats undercoat on primed wood general surfaces						
exceeding 300mm girth	m2	0.36	3.64	1.00	0.70	5.34
not exceeding 150mm girth	m	0.08	0.81	0.16	0.15	1.11
150-300mm girth	m	0.12	1.21	0.32	0.23	1.76
isolated areas not exceeding 0.5m2	nr	0.20	2.02	0.48	0.38	2.88

	Unit	Hours	Hours £	Materials £	O & P £	Total £
Two coats undercoat on primed wood windows, glazed doors and the like						
panes area not exceeding 0.1m2	m2	0.60	6.07	0.40	0.97	7.44
panes area 0.1 to 0.5m2	m2	0.50	5.06	0.36	0.81	6.23
panes area 0.5 to 1m2	m2	0.40	4.04	0.32	0.65	5.02
panes area exceeding 1m2	m2	0.30	3.03	0.28	0.50	3.81
Two coats undercoat on primed wood frames, linings, skirtings and the like						
exceeding 300mm girth	m2	0.40	4.04	1.00	0.76	5.80
not exceeding 150mm girth	m	0.10	1.01	0.16	0.18	1.35
150-300mm girth	m	0.16	1.62	0.32	0.29	2.23
isolated areas not exceeding 0.5m2	nr	0.22	2.23	0.48	0.41	3.12
One coat gloss on undercoated wood general surfaces						
exceeding 300mm girth	m2	0.18	1.82	0.50	0.35	2.67
not exceeding 150mm girth	m	0.04	0.40	0.08	0.07	0.56
150-300mm girth	m	0.06	0.61	0.16	0.11	0.88
isolated areas not exceeding 0.5m2	nr	0.10	1.01	0.24	0.19	1.44
One coat gloss on undercoated wood windows, glazed doors and the like						
panes area not exceeding 0.1m2	m2	0.35	3.54	0.20	0.56	4.30
panes area 0.1 to 0.5m2	m2	0.30	3.03	0.18	0.48	3.69
panes area 0.5 to 1m2	m2	0.25	2.53	0.16	0.40	3.09
panes area exceeding 1m2	m2	0.22	2.22	0.14	0.35	2.72

	Unit	Hours	Hours £	Materials £	O & P £	Total £
One coat gloss on undercoated wood frames, linings, skirtings and the like						
exceeding 300mm girth	m2	0.20	2.02	0.50	0.38	2.90
not exceeding 150mm girth	m	0.05	0.51	0.08	0.09	0.67
150-300mm girth	m	0.08	0.81	0.16	0.15	1.11
isolated areas not exceeding 0.5m2	nr	0.11	1.11	0.24	0.20	1.55
One coat primer, one coat undercoat and one coat gloss on wood general surfaces						
exceeding 300mm girth	m2	0.54	5.46	1.50	1.04	8.00
not exceeding 150mm girth	m	0.12	1.21	0.24	0.22	1.67
150-300mm girth	m	0.18	1.82	0.48	0.34	2.64
isolated areas not exceeding 0.5m2	nr	0.30	3.03	0.72	0.56	4.32
One coat primer, one coat undercoat and one coat gloss on wood windows, glazed doors and the like						
panes area not exceeding 0.1m2	m2	0.75	7.58	0.58	1.22	9.39
panes area 0.1 to 0.5m2	m2	0.60	6.07	0.44	0.98	7.48
panes area 0.5 to 1m2	m2	0.50	5.06	0.46	0.83	6.34
panes area exceeding 1m2	m2	0.40	4.04	0.40	0.67	5.11
One coat primer, one coat undercoat and one coat gloss on wood frames, linings, skirtings and the like						
exceeding 300mm girth	m2	0.60	6.07	1.46	1.13	8.65
not exceeding 150mm girth	m	0.15	1.52	0.22	0.26	2.00
150-300mm girth	m	0.24	2.43	0.46	0.43	3.32
isolated areas not exceeding 0.5m2	nr	0.33	3.34	0.70	0.61	4.64

	Unit	Hours	Hours £	Materials £	O & P £	Total £
One coat primer, two coats undercoat and one coat gloss on wood general surfaces						
exceeding 300mm girth	m2	0.72	7.28	2.00	1.39	10.67
not exceeding 150mm girth	m	0.16	1.62	0.32	0.29	2.23
150-300mm girth	m	0.24	2.43	0.64	0.46	3.53
isolated areas not exceeding 0.5m2	nr	0.40	4.04	0.96	0.75	5.75
One coat primer, two coats undercoat and one coat gloss on wood windows, glazed doors and the like						
panes area not exceeding 0.1m2	m2	1.00	10.11	0.78	1.63	12.52
panes area 0.1 to 0.5m2	m2	0.80	8.09	0.70	1.32	10.11
panes area 0.5 to 1m2	m2	0.70	7.08	0.62	1.15	8.85
panes area exceeding 1m2	m2	0.60	6.07	0.54	0.99	7.60
One coat primer, two coats undercoat and one coat gloss on wood frames, linings, skirtings and the like						
exceeding 300mm girth	m2	0.80	8.09	1.96	1.51	11.56
not exceeding 150mm girth	m	0.20	2.02	0.30	0.35	2.67
150-300mm girth	m	0.32	3.24	0.62	0.58	4.43
isolated areas not exceeding 0.5m2	nr	0.44	4.45	0.94	0.81	6.20

	Unit	Hours	Hours £	Materials £	O & P £	Total £
On internal metalwork						
One coat primer on metal general surfaces						
exceeding 300mm girth	m2	0.18	1.82	0.55	0.36	2.73
not exceeding 150mm girth	m	0.04	0.40	0.10	0.08	0.58
150-300mm girth	m	0.06	0.61	0.18	0.12	0.90
isolated areas not exceeding 0.5m2	nr	0.10	1.01	0.28	0.19	1.48
One coat primer on metal windows, glazed doors and the like						
panes area not exceeding 0.1m2	m2	0.35	3.54	0.20	0.56	4.30
panes area 0.1 to 0.5m2	m2	0.30	3.03	0.18	0.48	3.69
panes area 0.5 to 1m2	m2	0.25	2.53	0.16	0.40	3.09
panes area exceeding 1m2	m2	0.22	2.22	0.14	0.35	2.72
One coat primer on structural steelwork						
exceeding 300mm girth	m2	0.24	2.43	0.55	0.45	3.42
One coat primer on metal radiators						
exceeding 300mm girth	m2	0.18	1.82	0.55	0.36	2.73
One coat undercoat on metal general surfaces						
exceeding 300mm girth	m2	0.18	1.82	0.50	0.35	2.67
not exceeding 150mm girth	m	0.04	0.40	0.08	0.07	0.56
150-300mm girth	m	0.06	0.61	0.16	0.11	0.88
isolated areas not exceeding 0.5m2	nr	0.10	1.01	0.24	0.19	1.44

	Unit	Hours	Hours £	Materials £	O & P £	Total £
One coat undercoat on metal windows, glazed doors and the like						
panes area not exceeding 0.1m2	m2	0.35	3.54	0.18	0.56	4.28
panes area 0.1 to 0.5m2	m2	0.30	3.03	0.16	0.48	3.67
panes area 0.5 to 1m2	m2	0.25	2.53	0.14	0.40	3.07
panes area exceeding 1m2	m2	0.22	2.22	0.12	0.35	2.70
One coat undercoat on structural steelwork						
exceeding 300mm girth	m2	0.24	2.43	0.50	0.44	3.37
One coat undercoat on metal radiators						
exceeding 300mm girth	m2	0.18	1.82	0.50	0.35	2.67
Two coats undercoat on metal general surfaces						
exceeding 300mm girth	m2	0.18	1.82	1.00	0.42	3.24
not exceeding 150mm girth	m	0.04	0.40	0.16	0.08	0.65
150-300mm girth	m	0.06	0.61	0.32	0.14	1.07
isolated areas not exceeding 0.5m2	nr	0.10	1.01	0.48	0.22	1.71
Two coats undercoat on metal windows, glazed doors and the like						
panes area not exceeding 0.1m2	m2	0.35	3.54	0.36	0.58	4.48
panes area 0.1 to 0.5m2	m2	0.30	3.03	0.32	0.50	3.86
panes area 0.5 to 1m2	m2	0.25	2.53	0.28	0.42	3.23
panes area exceeding 1m2	m2	0.22	2.22	0.24	0.37	2.83

	Unit	Hours	Hours £	Materials £	O & P £	Total £
Two coats undercoat on structural steelwork						
exceeding 300mm girth	m2	0.24	2.43	1.00	0.51	3.94
Two coats undercoat on metal radiators						
exceeding 300mm girth	m2	0.18	1.82	1.00	0.42	3.24
One coat gloss on metal general surfaces						
exceeding 300mm girth	m2	0.18	1.82	0.50	0.35	2.67
not exceeding 150mm girth	m	0.04	0.40	0.08	0.07	0.56
150-300mm girth	m	0.06	0.61	0.16	0.11	0.88
isolated areas not exceeding 0.5m2	nr	0.10	1.01	0.24	0.19	1.44
One coat gloss on metal windows, glazed doors and the like						
panes area not exceeding 0.1m2	m2	0.35	3.54	0.18	0.56	4.28
panes area 0.1 to 0.5m2	m2	0.30	3.03	0.16	0.48	3.67
panes area 0.5 to 1m2	m2	0.25	2.53	0.14	0.40	3.07
panes area exceeding 1m2	m2	0.22	2.22	0.12	0.35	2.70
One coat gloss on structural steelwork						
exceeding 300mm girth	m2	0.24	2.43	0.50	0.44	3.37
One coat gloss on metal radiators						
exceeding 300mm girth	m2	0.18	1.82	0.50	0.35	2.67

	Unit	Hours	Hours £	Materials £	O & P £	Total £
One coat primer, one coat undercoat and one coat gloss on metal general surfaces						
exceeding 300mm girth	m2	0.54	5.46	1.55	1.05	8.06
not exceeding 150mm girth	m	0.12	1.21	0.26	0.22	1.69
150-300mm girth	m	0.18	1.82	0.50	0.35	2.67
isolated areas not exceeding 0.5m2	nr	0.30	3.03	0.76	0.57	4.36
One coat primer, one coat undercoat and one coat gloss on wood windows, glazed doors and the like						
panes area not exceeding 0.1m2	m2	0.75	7.58	0.60	1.23	9.41
panes area 0.1 to 0.5m2	m2	0.60	6.07	0.46	0.98	7.50
panes area 0.5 to 1m2	m2	0.50	5.06	0.48	0.83	6.37
panes area exceeding 1m2	m2	0.40	4.04	0.42	0.67	5.13
One coat primer, one coat undercoat and one coat gloss on structural steelwork						
exceeding 300mm girth	m2	0.60	6.07	1.55	1.14	8.76
One coat primer, one coat undercoat and one coat gloss on metal radiators						
exceeding 300mm girth	m2	0.52	5.26	1.55	1.02	7.83
One coat primer, two coats undercoat and one coat gloss on metal general surfaces						
exceeding 300mm girth	m2	0.54	5.46	2.05	1.13	8.64
not exceeding 150mm girth	m	0.12	1.21	0.26	0.22	1.69
150-300mm girth	m	0.18	1.82	0.50	0.35	2.67
isolated areas not exceeding .5m2	nr	0.30	3.03	0.74	0.57	4.34

	Unit	Hours	Hours £	Materials £	O & P £	Total £
One coat primer, two coats undercoat and one coat gloss on metal windows, glazed doors and the like						
panes area not exceeding 0.1m2	m2	0.75	7.58	0.74	1.25	9.57
panes area 0.1 to 0.5m2	m2	0.60	6.07	0.66	1.01	7.73
panes area 0.5 to 1m2	m2	0.50	5.06	0.60	0.85	6.50
panes area exceeding 1m2	m2	0.40	4.04	0.50	0.68	5.23
One coat primer, two coats undercoat and one coat gloss on structural steelwork						
exceeding 300mm girth	m2	0.78	7.89	2.05	1.49	11.43
One coat primer, two coats undercoat and one coat gloss on metal radiators						
exceeding 300mm girth	m2	0.72	7.28	2.01	1.39	10.68
On external woodwork						
One coat primer on wood general surfaces						
exceeding 300mm girth	m2	0.24	2.43	0.50	0.44	3.37
not exceeding 150mm girth	m	0.06	0.61	0.08	0.10	0.79
150-300mm girth	m	0.08	0.81	0.16	0.15	1.11
isolated areas not exceeding 0.5m2	nr	0.02	0.20	0.24	0.07	0.51
One coat primer on wood windows, glazed doors and the like						
panes area not exceeding 0.1m2	m2	0.40	4.04	0.18	0.63	4.86
panes area 0.1 to 0.5m2	m2	0.35	3.54	0.16	0.55	4.25
panes area 0.5 to 1m2	m2	0.30	3.03	0.14	0.48	3.65
panes area exceeding 1m2	m2	0.25	2.53	0.12	0.40	3.04

	Unit	Hours	Hours £	Materials £	O & P £	Total £
One coat primer on wood frames and the like						
exceeding 300mm girth	m2	0.26	2.63	0.46	0.46	3.55
not exceeding 150mm girth	m	0.07	0.71	0.06	0.12	0.88
150-300mm girth	m	0.10	1.01	0.14	0.17	1.32
isolated areas not exceeding 0.5m2	nr	0.14	1.42	0.22	0.25	1.88
One coat undercoat on primed wood general surfaces						
exceeding 300mm girth	m2	0.24	2.43	0.50	0.44	3.37
not exceeding 150mm girth	m	0.06	0.61	0.08	0.10	0.79
150-300mm girth	m	0.08	0.81	0.16	0.15	1.11
isolated areas not exceeding 0.5m2	nr	0.12	1.21	0.24	0.22	1.67
One coat undercoat on primed wood windows, glazed doors and the like						
panes area not exceeding 0.1m2	m2	0.40	4.04	0.20	0.64	4.88
panes area 0.1 to 0.5m2	m2	0.35	3.54	0.18	0.56	4.28
panes area 0.5 to 1m2	m2	0.30	3.03	0.16	0.48	3.67
panes area exceeding 1m2	m2	0.25	2.53	0.14	0.40	3.07
One coat undercoat on primed wood frames and the like						
exceeding 300mm girth	m2	0.26	2.63	0.50	0.47	3.60
not exceeding 150mm girth	m	0.07	0.71	0.08	0.12	0.91
150-300mm girth	m	0.10	1.01	0.16	0.18	1.35
isolated areas not exceeding 0.5m2	nr	0.14	1.42	0.24	0.25	1.90

	Unit	Hours	Hours £	Materials £	O & P £	Total £
Two coats undercoat on primed wood general surfaces						
exceeding 300mm girth	m2	0.48	4.85	1.00	0.88	6.73
not exceeding 150mm girth	m	0.12	1.21	0.16	0.21	1.58
150-300mm girth	m	0.16	1.62	0.32	0.29	2.23
isolated areas not exceeding 0.5m2	nr	0.24	2.43	0.48	0.44	3.34
Two coats undercoat on primed wood windows, glazed doors and the like						
panes area not exceeding 0.1m2	m2	0.80	8.09	0.40	1.27	9.76
panes area 0.1 to 0.5m2	m2	0.70	7.08	0.36	1.12	8.55
panes area 0.5 to 1m2	m2	0.60	6.07	0.32	0.96	7.34
panes area exceeding 1m2	m2	0.50	5.06	0.28	0.80	6.14
Two coats undercoat on primed wood frames and the like						
exceeding 300mm girth	m2	0.52	5.26	1.00	0.94	7.20
not exceeding 150mm girth	m	0.14	1.42	0.16	0.24	1.81
150-300mm girth	m	0.20	2.02	0.32	0.35	2.69
isolated areas not exceeding 0.5m2	nr	0.32	3.24	0.48	0.56	4.27
One coat gloss on undercoated wood general surfaces						
exceeding 300mm girth	m2	0.24	2.43	0.50	0.44	3.37
not exceeding 150mm girth	m	0.06	0.61	0.08	0.10	0.79
150-300mm girth	m	0.08	0.81	0.16	0.15	1.11
isolated areas not exceeding 0.5m2	nr	0.12	1.21	0.24	0.22	1.67

	Unit	Hours	Hours £	Materials £	O & P £	Total £
One coat gloss on undercoated wood windows, glazed doors and the like						
panes area not exceeding 0.1m2	m2	0.40	4.04	0.20	0.64	4.88
panes area 0.1 to 0.5m2	m2	0.35	3.54	0.18	0.56	4.28
panes area 0.5 to 1m2	m2	0.30	3.03	0.16	0.48	3.67
panes area exceeding 1m2	m2	0.25	2.53	0.14	0.40	3.07
One coat gloss on undercoated wood frames and the like						
exceeding 300mm girth	m2	0.26	2.63	0.50	0.47	3.60
not exceeding 150mm girth	m	0.07	0.71	0.08	0.12	0.91
150-300mm girth	m	0.10	1.01	0.16	0.18	1.35
isolated areas not exceeding 0.5m2	nr	0.16	1.62	0.24	0.28	2.14
One coat primer, one coat undercoat and one coat gloss on wood general surfaces						
exceeding 300mm girth	m2	0.73	7.33	1.50	1.32	10.15
not exceeding 150mm girth	m	0.14	1.42	0.24	0.25	1.90
150-300mm girth	m	0.24	2.43	0.48	0.44	3.34
isolated areas not exceeding 0.5m2	nr	0.40	4.04	0.72	0.71	5.48
One coat primer, one coat undercoat and one coat gloss on wood windows, glazed doors and the like						
panes area not exceeding 0.1m2	m2	1.20	12.13	0.58	1.91	14.62
panes area 0.1 to 0.5m2	m2	1.05	10.62	0.44	1.66	12.71
panes area 0.5 to 1m2	m2	0.90	9.10	0.46	1.43	10.99
panes area exceeding 1m2	m2	0.75	7.58	0.40	1.20	9.18

	Unit	Hours	Hours £	Materials £	O & P £	Total £
One coat primer, one coat undercoat and one coat gloss on wood frames and the like						
exceeding 300mm girth	m2	0.78	7.89	1.46	1.40	10.75
not exceeding 150mm girth	m	0.21	2.12	0.22	0.35	2.69
150-300mm girth	m	0.30	3.03	0.46	0.52	4.02
isolated areas not exceeding 0.5m2	nr	0.48	4.85	0.70	0.83	6.39
One coat primer, two coats undercoat and one coat gloss on wood general surfaces						
exceeding 300mm girth	m2	0.96	9.71	2.00	1.76	13.46
not exceeding 150mm girth	m	0.25	2.53	0.32	0.43	3.27
150-300mm girth	m	0.34	3.44	0.64	0.61	4.69
isolated areas not exceeding 0.5m2	nr	0.48	4.85	0.96	0.87	6.68
One coat primer, two coats undercoat and one coat gloss on wood windows, glazed doors and the like						
panes area not exceeding 0.1m2	m2	1.60	16.18	0.78	2.54	19.50
panes area 0.1 to 0.5m2	m2	1.40	14.15	0.70	2.23	17.08
panes area 0.5 to 1m2	m2	1.20	12.13	0.62	1.91	14.66
panes area exceeding 1m2	m2	1.00	10.11	0.54	1.60	12.25
One coat primer, two coats undercoat and one coat gloss on wood frames and the like						
exceeding 300mm girth	m2	1.04	10.51	1.96	1.87	14.35
not exceeding 150mm girth	m	0.28	2.83	0.30	0.47	3.60
150-300mm girth	m	0.40	4.04	0.62	0.70	5.36
isolated areas not exceeding .5m2	nr	0.64	6.47	0.94	1.11	8.52

	Unit	Hours	Hours £	Materials £	O & P £	Total £

On external metalwork

One coat primer on metal general surfaces

exceeding 300mm girth	m2	0.24	2.43	0.55	0.45	3.42
not exceeding 150mm girth	m	0.06	0.61	0.10	0.11	0.81
150-300mm girth	m	0.08	0.81	0.18	0.15	1.14
isolated areas not exceeding 0.5m2	nr	0.12	1.21	0.28	0.22	1.72

One coat primer on metal windows, glazed doors and the like

panes area not exceeding 0.1m2	m2	0.40	4.04	0.20	0.64	4.88
panes area 0.1 to 0.5m2	m2	0.35	3.54	0.18	0.56	4.28
panes area 0.5 to 1m2	m2	0.30	3.03	0.16	0.48	3.67
panes area exceeding 1m2	m2	0.25	2.53	0.14	0.40	3.07

One coat primer on structural steelwork

exceeding 300mm girth	m2	0.30	3.03	0.55	0.54	4.12

One coat primer on pipes and gutters

exceeding 300mm girth	m2	0.26	2.63	0.55	0.48	3.66
150-300mm girth	m	0.08	0.81	0.18	0.15	1.14

One coat undercoat on metal general surfaces

exceeding 300mm girth	m2	0.24	2.43	0.50	0.44	3.37
not exceeding 150mm girth	m	0.06	0.61	0.08	0.10	0.79
150-300mm girth	m	0.08	0.81	0.16	0.15	1.11
isolated areas not exceeding 0.5m2	nr	0.12	1.21	0.24	0.22	1.67

	Unit	Hours	Hours £	Materials £	O & P £	Total £
One coat undercoat on metal windows, glazed doors and the like						
panes area not exceeding 0.1m2	m2	0.40	4.04	0.18	0.63	4.86
panes area 0.1 to 0.5m2	m2	0.35	3.54	0.16	0.55	4.25
panes area 0.5 to 1m2	m2	0.30	3.03	0.14	0.48	3.65
panes area exceeding 1m2	m2	0.25	2.53	0.12	0.40	3.04
One coat undercoat on structural steelwork						
exceeding 300mm girth	m2	0.30	3.03	0.50	0.53	4.06
One coat primer on pipes and gutters						
exceeding 300mm girth	m2	0.26	2.63	0.50	0.47	3.60
150-300mm girth	m	0.08	0.81	0.16	0.15	1.11
Two coats undercoat on metal general surfaces						
exceeding 300mm girth	m2	0.48	4.85	1.00	0.88	6.73
not exceeding 150mm girth	m	0.12	1.21	0.16	0.21	1.58
150-300mm girth	m	0.20	2.02	0.32	0.35	2.69
isolated areas not exceeding 0.5m2	nr	0.24	2.43	0.48	0.44	3.34
Two coats undercoat on metal windows, glazed doors and the like						
panes area not exceeding 0.1m2	m2	0.80	8.09	0.36	1.27	9.72
panes area 0.1 to 0.5m2	m2	0.70	7.08	0.32	1.11	8.51
panes area 0.5 to 1m2	m2	0.60	6.07	0.28	0.95	7.30
panes area exceeding 1m2	m2	0.50	5.06	0.24	0.79	6.09

	Unit	Hours	Hours £	Materials £	O & P £	Total £
Two coats undercoat on structural steelwork						
exceeding 300mm girth	m2	0.60	6.07	1.00	1.06	8.13
Two coats undercoat on pipes and gutters						
exceeding 300mm girth	m2	0.52	5.26	1.00	0.94	7.20
150-300mm girth	m	0.16	1.62	0.32	0.29	2.23
One coat gloss on metal general surfaces						
exceeding 300mm girth	m2	0.24	2.43	0.50	0.44	3.37
not exceeding 150mm girth	m	0.06	0.61	0.08	0.10	0.79
150-300mm girth	m	0.08	0.81	0.16	0.15	1.11
isolated areas not exceeding 0.5m2	nr	0.12	1.21	0.24	0.22	1.67
One coat gloss on metal windows, glazed doors and the like						
panes area not exceeding 0.1m2	m2	0.40	4.04	0.18	0.63	4.86
panes area 0.1 to 0.5m2	m2	0.35	3.54	0.16	0.55	4.25
panes area 0.5 to 1m2	m2	0.30	3.03	0.14	0.48	3.65
panes area exceeding 1m2	m2	0.25	2.53	0.12	0.40	3.04
One coat gloss on structural steelwork						
exceeding 300mm girth	m2	0.30	3.03	0.50	0.53	4.06
One coat gloss on pipes and gutters						
exceeding 300mm girth	m2	0.26	2.63	0.50	0.47	3.60
150-300mm girth	m	0.08	0.81	0.18	0.15	1.14

	Unit	Hours	Hours £	Materials £	O & P £	Total £
One coat primer, one coat undercoat and one coat gloss on metal general surfaces						
exceeding 300mm girth	m2	0.72	7.28	1.55	1.32	10.15
not exceeding 150mm girth	m	0.18	1.82	0.26	0.31	2.39
150-300mm girth	m	0.24	2.43	0.50	0.44	3.37
isolated areas not exceeding 0.5m2	nr	0.36	3.64	0.76	0.66	5.06
One coat primer, one coat undercoat and one coat gloss on metal windows, glazed doors and the like						
panes area not exceeding 0.1m2	m2	1.20	12.13	0.60	1.91	14.64
panes area 0.1 to 0.5m2	m2	1.05	10.62	0.46	1.66	12.74
panes area 0.5 to 1m2	m2	0.90	9.10	0.48	1.44	11.02
panes area exceeding 1m2	m2	0.75	7.58	0.42	1.20	9.20
One coat primer, one coat undercoat and one coat gloss on structural steelwork						
exceeding 300mm girth	m2	0.90	9.10	1.55	1.60	12.25
One coat primer, one coat undercoat and one coat gloss on pipes and gutters						
exceeding 300mm girth	m2	0.78	7.89	1.50	1.41	10.79
150-300mm girth	m	0.24	2.43	0.54	0.44	3.41

	Unit	Hours	Hours £	Materials £	O & P £	Total £
One coat primer, two coats undercoat and one coat gloss on metal general surfaces						
exceeding 300mm girth	m2	0.96	9.71	2.05	1.76	13.52
not exceeding 150mm girth	m	0.24	2.43	0.26	0.40	3.09
150-300mm girth	m	0.32	3.24	0.50	0.56	4.30
isolated areas not exceeding 0.5m2	nr	0.48	4.85	0.74	0.84	6.43
One coat primer, two coats undercoat and one coat gloss on metal windows, glazed doors and the like						
panes area not exceeding 0.1m2	m2	1.60	16.18	0.74	2.54	19.45
panes area 0.1 to 0.5m2	m2	1.40	14.15	0.66	2.22	17.04
panes area 0.5 to 1m2	m2	1.20	12.13	0.60	1.91	14.64
panes area exceeding 1m2	m2	1.00	10.11	0.50	1.59	12.20
One coat primer, two coats undercoat and one coat gloss on structural steelwork						
exceeding 300mm girth	m2	1.20	12.13	2.05	2.13	16.31
One coat primer, two coats undercoat and one coat gloss on pipes and gutters						
exceeding 300mm girth	m2	1.04	10.51	2.00	1.88	14.39
150-300mm girth	m	0.32	3.24	0.76	0.60	4.59

	Unit	Hours	Hours £	Materials £	O & P £	Total £

Wallpapering

Prepare, size, apply adhesive, supply
and hang paper to plastered walls
and columns, butt jointed

	Unit	Hours	Hours £	Materials £	O & P £	Total £
lining paper						
£1.50 per roll	m2	0.30	3.03	0.36	0.51	3.90
£2.00 per roll	m2	0.30	3.03	0.48	0.53	4.04
£2.50 per roll	m2	0.30	3.03	0.59	0.54	4.17
washable paper						
£2.50 per roll	m2	0.30	3.03	0.59	0.54	4.17
£4.00 per roll	m2	0.30	3.03	0.94	0.60	4.57
£5.00 per roll	m2	0.30	3.03	1.18	0.63	4.84
vinyl paper						
£4.00 per roll	m2	0.30	3.03	0.94	0.60	4.57
£5.00 per roll	m2	0.30	3.03	1.18	0.63	4.84
£6.00 per roll	m2	0.30	3.03	1.41	0.67	5.11
washable paper						
£5.00 per roll	m2	0.30	3.03	1.18	0.63	4.84
£6.00 per roll	m2	0.30	3.03	1.41	0.67	5.11
£7.00 per roll	m2	0.30	3.03	1.64	0.70	5.37
hessian paper						
£7.00 per m2	m2	0.50	5.06	7.70	1.91	14.67
£8.00 per m2	m2	0.50	5.06	8.80	2.08	15.93
£9.00 per m2	m2	0.50	5.06	9.90	2.24	17.20
suede paper						
£9.00 per m2	m2	0.30	3.03	9.90	1.94	14.87
£10.00 per m2	m2	0.30	3.03	11.00	2.10	16.14
£11.00 per m2	m2	0.30	3.03	12.10	2.27	17.40

	Unit	Hours	Hours £	Materials £	O & P £	Total £

Prepare, size, apply adhesive, supply
and hang paper to plastered ceilings
and columns, butt jointed

	Unit	Hours	Hours £	Materials £	O & P £	Total £
lining paper						
£1.50 per roll	m2	0.35	3.54	0.36	0.58	4.48
£2.00 per roll	m2	0.35	3.54	0.48	0.60	4.62
£2.50 per roll	m2	0.35	3.54	0.59	0.62	4.75
washable paper						
£2.50 per roll	m2	0.35	3.54	0.59	0.62	4.75
£4.00 per roll	m2	0.35	3.54	0.94	0.67	5.15
£5.00 per roll	m2	0.35	3.54	1.18	0.71	5.43
vinyl paper						
£4.00 per roll	m2	0.35	3.54	0.94	0.67	5.15
£5.00 per roll	m2	0.35	3.54	1.18	0.71	5.43
£6.00 per roll	m2	0.35	3.54	1.41	0.74	5.69
washable paper						
£5.00 per roll	m2	0.35	3.54	1.18	0.71	5.43
£6.00 per roll	m2	0.35	3.54	1.41	0.74	5.69
£7.00 per roll	m2	0.35	3.54	1.64	0.78	5.96

	Unit	Hours	Hours £	Materials £	O & P £	Total £

EXTERNAL WORKS

Drainage

Excavate drain trench by hand, support sides, level and ram bottom of trench, backfill and consolidate with excavated material and remove surplus to skip for pipes 100mm and 150mm diameter, average depth of trench

	Unit	Hours	Hours £	Materials £	O & P £	Total £
0.50m	m	1.20	9.60	0.00	1.44	11.04
0.75m	m	1.90	15.20	0.00	2.28	17.48
1.00m	m	2.50	20.00	0.00	3.00	23.00
1.25m	m	4.00	32.00	0.00	4.80	36.80
1.50m	m	5.00	40.00	0.00	6.00	46.00
1.75m	m	6.00	48.00	0.00	7.20	55.20
2.00m	m	7.20	57.60	0.00	8.64	66.24
2.25m	m	8.20	65.60	0.00	9.84	75.44
2.50m	m	9.50	76.00	0.00	11.40	87.40
2.75m	m	10.50	84.00	0.00	12.60	96.60
3.00m	m	12.00	96.00	0.00	14.40	110.40

Excavate drain trench by hand, support sides, level and ram bottom of trench, backfill and consolidate with excavated material and remove surplus to skip for pipes 225mm diameter, average depth of trench

	Unit	Hours	Hours £	Materials £	O & P £	Total £
0.50m	m	1.30	10.40	0.00	1.56	11.96
0.75m	m	2.10	16.80	0.00	2.52	19.32
1.00m	m	2.70	21.60	0.00	3.24	24.84
1.25m	m	4.30	34.40	0.00	5.16	39.56
1.50m	m	5.40	43.20	0.00	6.48	49.68
1.75m	m	6.60	52.80	0.00	7.92	60.72

	Unit	Hours	Hours £	Materials £	O & P £	Total £
2.00m	m	7.10	56.80	0.00	8.52	65.32
2.25m	m	9.00	72.00	0.00	10.80	82.80
2.50m	m	10.20	81.60	0.00	12.24	93.84
2.75m	m	11.30	90.40	0.00	13.56	103.96
3.00m	m	12.80	102.40	0.00	15.36	117.76

Extra for breaking up by hand

	Unit	Hours	Hours £	Materials £	O & P £	Total £
plain concrete 100mm thick	m2	0.90	7.20	0.00	1.08	8.28
reinforced concrete 100mm thick	m2	1.00	8.00	0.00	1.20	9.20
tarmacadam 75mm thick	m2	0.55	4.40	0.00	0.66	5.06
hardcore 100mm thick	m2	0.40	3.20	0.00	0.48	3.68
soft rock	m3	8.00	64.00	0.00	9.60	73.60
hard rock	m3	10.00	80.00	0.00	12.00	92.00

	Unit	Hours	Hours £	Plant £	O & P £	Total £

Excavate drain trench by machine, support sides, level and ram bottom of trench, backfill and consolidate with excavated material and remove surplus to skip for pipes 100mm and 150mm diameter, average depth of trench

	Unit	Hours	Hours £	Plant £	O & P £	Total £
0.50m	m	0.30	2.40	1.80	0.63	4.83
0.75m	m	0.50	4.00	2.70	1.01	7.71
1.00m	m	0.90	7.20	4.00	1.68	12.88
1.25m	m	1.50	12.00	5.50	2.63	20.13
1.50m	m	1.80	14.40	6.00	3.06	23.46
1.75m	m	2.20	17.60	7.00	3.69	28.29
2.00m	m	2.80	22.40	8.00	4.56	34.96
2.25m	m	3.40	27.20	10.00	5.58	42.78
2.50m	m	3.90	31.20	11.50	6.41	49.11
2.75m	m	4.50	36.00	13.00	7.35	56.35
3.00m	m	5.00	40.00	14.50	8.18	62.68

	Unit	Hours	Hours £	Plant £	O & P £	Total £

Excavate drain trench by machine, support sides, level and ram bottom of trench, backfill and consolidate with excavated material and remove surplus to skip for pipes 225mm diameter, average depth of trench

	Unit	Hours	Hours £	Plant £	O & P £	Total £
0.50m	m	1.30	10.40	1.80	1.83	14.03
0.75m	m	2.10	16.80	2.70	2.93	22.43
1.0m	m	2.70	21.60	4.10	3.86	29.56
1.25m	m	4.30	34.40	5.75	6.02	46.17
1.50m	m	5.40	43.20	6.25	7.42	56.87
1.75m	m	6.60	52.80	7.40	9.03	69.23
2.00m	m	7.10	56.80	8.55	9.80	75.15
2.25m	m	9.00	72.00	10.20	12.33	94.53
2.50m	m	10.20	81.60	11.50	13.97	107.07
2.75m	m	11.30	90.40	13.50	15.59	119.49
3.00m	m	12.80	102.40	14.75	17.57	134.72

Extra for breaking up by machine

	Unit	Hours	Hours £	Plant £	O & P £	Total £
plain concrete 100mm thick	m2	0.30	2.40	1.50	0.59	4.49
reinforced concrete 100mm thick	m2	0.35	2.80	1.80	0.69	5.29
tarmacadam 75mm thick	m2	0.20	1.60	1.10	0.41	3.11
hardcore 100mm thick	m2	0.15	1.20	0.80	0.30	2.30
soft rock	m3	4.55	36.40	5.80	6.33	48.53
hard rock	m3	5.20	41.60	7.25	7.33	56.18

Sand bed in trench under pipe 100mm diameter, thickness

	Unit	Hours	Hours £	Plant £	O & P £	Total £
100mm	m	0.10	0.80	0.95	0.26	2.01
150mm	m	0.12	0.96	1.40	0.35	2.71

	Unit	Hours	Hours £	Materials £	O & P £	Total £
Sand bed in trench under pipe 150mm diameter, thickness						
100mm	m	0.11	0.88	1.30	0.33	2.51
150mm	m	0.13	1.04	1.45	0.37	2.86
Sand bed in trench under pipe 225mm diameter, thickness						
100mm	m	0.14	1.12	1.55	0.40	3.07
150mm	m	0.16	1.28	2.25	0.53	4.06
Granular filling in bed in trench under pipe 100mm diameter, thickness						
100mm	m	0.12	0.96	0.80	0.26	2.02
150mm	m	0.14	1.12	1.15	0.34	2.61
Granular filling in bed in trench under pipe 150mm diameter, thickness						
100mm	m	0.13	1.04	0.90	0.29	2.23
150mm	m	0.15	1.20	1.35	0.38	2.93
Granular filling in bed in trench under pipe 225mm diameter, thickness						
100mm	m	0.16	1.28	1.15	0.36	2.79
150mm	m	0.18	1.44	1.75	0.48	3.67
Concrete in bed in trench under pipe, 100mm diameter, thickness						
100mm	m	0.24	1.92	3.50	0.81	6.23
150mm	m	0.28	2.24	5.20	1.12	8.56

	Unit	Hours	Hours £	Materials £	O & P £	Total £
Concrete in bed in trench under pipe, 150mm diameter, thickness						
100mm	m	0.26	2.08	4.05	0.92	7.05
150mm	m	0.30	2.40	5.15	1.13	8.68
Concrete in bed in trench under pipe, 225mm diameter, thickness						
100mm	m	0.32	2.56	4.60	1.07	8.23
150mm	m	0.36	2.88	7.05	1.49	11.42
Concrete in bed and haunching to pipe, 100mm diameter, bed thickness						
100mm	m	0.48	3.84	6.40	1.54	11.78
150mm	m	0.56	4.48	9.56	2.11	16.15
Concrete in bed and haunching to pipe, 150mm diameter, bed thickness						
100mm	m	0.52	4.16	7.23	1.71	13.10
150mm	m	0.60	4.80	9.88	2.20	16.88
Concrete in bed and haunching to pipe, 225mm diameter, bed thickness						
100mm	m	0.60	4.80	8.14	1.94	14.88
150mm	m	0.66	5.28	10.54	2.37	18.19
Granular filling in bed and surround to pipe, 100mm diameter, thickness						
100mm	m	0.36	2.88	3.21	0.91	7.00
150mm	m	0.42	3.36	4.96	1.25	9.57

	Unit	Hours	Hours £	Materials £	O & P £	Total £
Granular filling in bed and surround to pipe, 150mm diameter, thickness						
100mm	m	0.40	3.20	4.53	1.16	8.89
150mm	m	0.46	3.68	6.85	1.58	12.11
Granular filling in bed and surround to pipe, 225mm diameter, thickness						
100mm	m	0.44	3.52	6.80	1.55	11.87
150mm	m	0.50	4.00	8.12	1.82	13.94
Concrete in bed and surround to pipe, 100mm diameter, thickness						
100mm	m	0.72	5.76	7.60	2.00	15.36
150mm	m	0.84	6.72	8.27	2.25	17.24
Concrete in bed and surround to pipe, 150mm diameter, thickness						
100mm	m	0.72	5.76	8.26	2.10	16.12
150mm	m	0.84	6.72	9.02	2.36	18.10
Concrete in bed and surround to pipe, 150mm diameter, thickness						
100mm	m	0.72	5.76	10.68	2.47	18.91
150mm	m	0.84	6.72	12.58	2.90	22.20

	Unit	Hours	Hours £	Materials £	O & P £	Total £
Hepworths' Supersleve vitrified clay drain pipes, spigot and socket joints with sealing rings, 100mm diameter						
laid in trenches	m	0.36	2.88	5.76	1.30	9.94
in lengths not exceeding 3m	m	0.54	4.32	5.76	1.51	11.59
bend	nr	0.30	2.40	7.84	1.54	11.78
rest bend	nr	0.30	2.40	12.19	2.19	16.78
single junction	nr	0.30	2.40	11.50	2.09	15.99
Hepworths' Supersleve vitrified clay drain pipes, spigot and socket joints with sealing rings, 150mm diameter						
laid in trenches	m	0.60	4.80	10.99	2.37	18.16
in lengths not exceeding 3m	m	0.40	3.20	10.99	2.13	16.32
bend	nr	0.30	2.40	10.96	2.00	15.36
rest bend	nr	0.30	2.40	14.09	2.47	18.96
single junction	nr	0.30	2.40	14.68	2.56	19.64
Vitrified clay inlet gulley complete with grid and surrounded in concrete	nr	1.50	12.00	36.56	7.28	55.84
Vitrified clay back inlet gulley complete with grid and surrounded in concrete	nr	1.50	12.00	52.25	9.64	73.89
Vitrified clay paved area gulley complete with grid and surrounded in concrete	nr	1.50	12.00	42.50	8.18	62.68

	Unit	Hours	Hours £	Materials £	O & P £	Total £

Manholes

Excavate by hand for manhole
not exceeding

1.0m deep	m3	4.00	32.00	0.00	4.80	36.80
1.5m deep	m3	4.50	36.00	0.00	5.40	41.40
2.0m deep	m3	5.60	44.80	0.00	6.72	51.52

Dispose surplus excavated material
by hand, load into barrows, wheel

50m and deposit into skip	m3	2.20	17.60	0.00	2.64	20.24

Excavate by machine for manhole
not exceeding

1.0m deep	m3	0.35	2.80	4.20	1.05	8.05
1.5m deep	m3	0.40	3.20	4.40	1.14	8.74
2.0m deep	m3	0.45	3.60	4.60	1.23	9.43

Earthwork support not exceeding
2m between opposing faces,
depth not exceeding

1.0m deep	m2	0.35	2.80	1.30	0.62	4.72
2.0m deep	m2	0.40	3.20	1.40	0.69	5.29
4.0m deep	m2	0.45	3.60	1.50	0.77	5.87

Site-mixed concrete in base of
manhole, thickness

100-150mm	m3	2.00	16.00	62.15	11.72	89.87
150-300mm	m3	1.80	14.40	62.15	11.48	88.03

Site-mixed concrete in benching
to manhole, average thickness

225mm	m3	6.00	48.00	69.47	17.62	135.09

	Unit	Hours	Hours £	Materials £	O & P £	Total £
Engineering bricks Class 'B' in cement mortar in walls of manhole	m2	4.80	38.40	49.90	13.25	101.55
Extra for fair face and flush pointing	m2	0.25	2.00	0.00	0.30	2.30
Build in ends of pipes to one brick wall and make good, pipe diameter						
100mm	nr	0.15	1.20	0.00	0.18	1.38
150mm	nr	0.19	1.51	0.00	0.23	1.74
Galvanised step iron built into brickwork	nr	0.20	1.60	5.37	1.05	8.02
Cast iron manhole cover, frame bedded in cement mortar						
Grade A, light duty, size 600 x 450mm	nr	1.90	15.20	58.98	11.13	85.31
Grade B, medium duty, size 600 x 450mm	nr	2.00	16.00	102.37	17.76	136.13
Best quality vitrified clay channels, bedded in cement mortar						
half-section straight main channel						
100mm diameter x 300mm	nr	0.75	6.00	2.65	1.30	9.95
100mm diameter x 600mm	nr	0.85	6.80	2.65	1.42	10.87
100mm diameter x 1000mm	nr	0.95	7.60	3.71	1.70	13.01
150mm diameter x 300mm	nr	0.75	6.00	4.46	1.57	12.03
150mm diameter x 600mm	nr	0.85	6.80	4.46	1.69	12.95
150mm diameter x 1000mm	nr	0.95	7.60	6.53	2.12	16.25
Half-section 90 degrees channel bend						
100mm	nr	0.30	2.40	3.76	0.92	7.08
150mm	nr	0.40	3.20	6.49	1.45	11.14

	Unit	Hours	Hours £	Materials £	O & P £	Total £
Three-quarter section 90 degrees channel bend						
100mm	nr	1.20	9.60	10.60	3.03	23.23
150mm	nr	1.30	10.40	18.44	4.33	33.17

			Unit	Specialist price £	O & P £	Total £
Fencing						
Chainlink fencing, galvanised steel mesh on three strained line wires fixed to concrete posts at 3000mm centres						
900mm high			m	12.80	1.92	14.72
1200mm high			m	15.66	2.35	18.01
1400mm high			m	17.61	2.64	20.25
1800mm high			m	23.47	3.52	26.99
Chainlink fencing, galvanised steel mesh on three strained line wires fixed to galvanised mild steel posts at 3000mm centres						
900mm high			m	13.75	2.06	15.81
1200mm high			m	16.27	2.44	18.71
1400mm high			m	19.20	2.88	22.08
1800mm high			m	24.98	3.75	28.73
Chainlink fencing, plastic-coated mild steel mesh on three strained line wires fixed to concrete posts at 3000mm centres						
900mm high			m	13.35	2.00	15.35
1200mm high			m	16.41	2.46	18.87
1400mm high			m	18.36	2.75	21.11
1800mm high			m	24.22	3.63	27.85

	Unit	Specialist price £	O & P £	Total £
Chainlink fencing, plastic-coated mild steel mesh on three strained line wires fixed to galvanised mild steel posts at 3000mm centres				
900mm high	m	14.50	2.18	16.68
1200mm high	m	17.02	2.55	19.57
1400mm high	m	19.95	2.99	22.94
1800mm high	m	25.73	3.86	29.59
Strained wire fence including concrete posts at 3000mm centres				
1.00m high, 5 wires	m	10.88	1.63	12.51
1.20m high, 6 wires	m	11.18	1.68	12.86
1.40m high, 8 wires	m	11.78	1.77	13.55
Cleft chestnut fencing, pales set 75mm apart, 75mm diameter softwood posts at 2500mm centres				
0.90m high, 2 wires	m	9.80	1.47	11.27
1.05m high, 2 wires	m	10.05	1.51	11.56
1.20m high, 2 wires	m	10.56	1.58	12.14
1.50m high, 3 wires	m	11.54	1.73	13.27
1.80m high, 4 wires	m	12.47	1.87	14.34
Close-boarded fencing consisting of 90 x 19mm pales lapped 13mm fixed to 75 x 38mm horizontal rails on 75 x 75mm posts at 3000mm centres, height				
1.00m	m	31.25	4.69	35.94
1.20m	m	32.75	4.91	37.66
1.40m	m	34.25	5.14	39.39
1.60m	m	35.75	5.36	41.11
1.80m	m	37.25	5.59	42.84

	Unit	Hours	Hours £	Materials £	O & P £	Total £

Kerbs and edgings

Excavate trench by hand for kerb foundation

	Unit	Hours	Hours £	Materials £	O & P £	Total £
200 x 75mm	m	0.10	0.80	0.00	0.12	0.92
250 x 100mm	m	0.12	0.96	0.00	0.14	1.10
300 x 100mm	m	0.14	1.12	0.00	0.17	1.29
450 x 150mm	m	0.20	1.60	0.00	0.24	1.84
600 x 200mm	m	0.40	3.20	0.00	0.48	3.68

Excavate curved trench by hand for kerb foundation

	Unit	Hours	Hours £	Materials £	O & P £	Total £
200 x 75mm	m	0.12	0.96	0.00	0.14	1.10
250 x 100mm	m	0.14	1.12	0.00	0.17	1.29
300 x 100mm	m	0.16	1.28	0.00	0.19	1.47
450 x 150mm	m	0.24	1.92	0.00	0.29	2.21
600 x 200mm	m	0.44	3.52	0.00	0.53	4.05

	Unit	Hours	Hours £	Plant £	O & P £	Total £

Excavate trench by machine for kerb foundation

	Unit	Hours	Hours £	Plant £	O & P £	Total £
200 x 75mm	m	0.00	0.00	0.65	0.10	0.75
250 x 100mm	m	0.00	0.00	0.70	0.11	0.81
300 x 100mm	m	0.00	0.00	0.75	0.11	0.86
450 x 150mm	m	0.00	0.00	1.23	0.18	1.41
600 x 200mm	m	0.00	0.00	2.20	0.33	2.53

Excavate curved trench by machine for kerb foundation

	Unit	Hours	Hours £	Plant £	O & P £	Total £
200 x 75mm	m	0.00	0.00	0.75	0.11	0.86
250 x 100mm	m	0.00	0.00	0.80	0.12	0.92
300 x 100mm	m	0.00	0.00	0.85	0.13	0.98
450 x 150mm	m	0.00	0.00	1.46	0.22	1.68
600 x 200mm	m	0.00	0.00	2.57	0.39	2.96

	Unit	Hours	Hours £	Materials £	O & P £	Total £
Site-mixed concrete in foundation for kerb size						
200 x 75mm	m	0.03	0.24	0.94	0.18	1.36
250 x 100mm	m	0.04	0.32	1.55	0.28	2.15
300 x 100mm	m	0.05	0.40	1.86	0.34	2.60
450 x 150mm	m	0.18	1.44	4.20	0.85	6.49
600 x 200mm	m	0.30	2.40	7.45	1.48	11.33
Precast concrete kerbs, channels and edgings, jointed and pointed in cement mortar						
kerbs, straight						
127 x 254mm	m	0.40	3.20	4.98	1.23	9.41
152 x 305mm	m	0.45	3.60	6.88	1.57	12.05
kerbs, curved						
127 x 254mm	m	0.50	4.00	6.65	1.60	12.25
152 x 305mm	m	0.55	4.40	8.47	1.93	14.80
channels, straight						
127 x 254mm	m	0.40	3.20	4.98	1.23	9.41
channels, curved						
127 x 254mm	m	0.50	4.00	6.65	1.60	12.25
edgings, straight						
51 x 152mm	m	0.30	2.40	2.84	0.79	6.03
51 x 203mm	m	0.30	2.40	3.32	0.86	6.58

Sub-bases

Beds and bases compacting in layers and grading

	Unit	Hours	Hours £	Materials £	O & P £	Total £
average thickness, 100mm						
granular fill	m2	0.10	0.80	2.60	0.51	3.91
sand	m2	0.12	0.96	2.40	0.50	3.86
hardcore	m2	0.13	1.04	2.30	0.50	3.84

	Unit	Hours	Hours £	Materials £	O & P £	Total £
average thickness, 150mm						
granular fill	m2	0.12	0.96	3.24	0.63	4.83
sand	m2	0.14	1.12	3.10	0.63	4.85
hardcore	m2	0.16	1.28	2.95	0.63	4.86
average thickness, 200mm						
granular fill	m2	0.14	1.12	4.22	0.80	6.14
sand	m2	0.16	1.28	3.80	0.76	5.84
hardcore	m2	0.18	1.44	3.65	0.76	5.85

Concrete beds

Site-mixed concrete in beds

	Unit	Hours	Hours £	Materials £	O & P £	Total £
not exceeding 150mm thick	m3	2.80	22.40	69.47	13.78	105.65
150 to 450mm thick	m3	2.20	17.60	69.47	13.06	100.13

Formwork to sides of concrete bases

	Unit	Hours	Hours £	Materials £	O & P £	Total £
not exceeding 250mm wide	m	0.65	5.20	1.62	1.02	7.84
250 to 500mm wide	m	0.90	7.20	2.64	1.48	11.32

Steel fabric reinforcement laid in concrete beds

	Unit	Hours	Hours £	Materials £	O & P £	Total £
ref A142, 2.22kg/m2	m2	0.15	1.20	1.24	0.37	2.81
ref A193, 3.02kg/m2	m2	0.18	1.44	1.49	0.44	3.37

Expansion joint, impregnated fibre-base joint filler, formed joint

	Unit	Hours	Hours £	Materials £	O & P £	Total £
12.5mm thick						
not exceeding 150mm wide	m	0.18	1.44	2.10	0.53	4.07
150 to 300mm wide	m	0.24	1.92	3.23	0.77	5.92
300 to 450mm wide	m	0.30	2.40	4.87	1.09	8.36

	Unit	Hours	Hours £	Materials £	O & P £	Total £
25mm thick						
not exceeding 150mm wide	m	0.20	1.60	2.87	0.67	5.14
150 to 300mm wide	m	0.26	2.08	4.27	0.95	7.30
300 to 450mm wide	m	0.32	2.56	6.63	1.02	10.21
Treat surfaces of concrete before setting						
tamping	m2	0.06	0.48	0.00	0.07	0.55
floating	m2	0.10	0.80	0.00	0.12	0.92
trowelling	m2	0.15	1.20	0.00	0.18	1.38

	Unit	Specialist price £	O & P £	Total £
Pavings				
Bitumen macadam paving consisting of 50mm base course with 28mm aggregate and 20mm thick wearing course with 6mm aggregate	m2	18.97	2.85	21.82
Bitumen macadam paving consisting of 70mm base course with 40mm aggregate and 25mm thick wearing course with 10mm aggregate	m2	22.66	3.40	26.06

	Unit	Hours	Hours £	Materials £	O & P £	Total £
Gravel paving in two layers, bottom layer clinker aggregate, second layer fine gravel aggregate						
50mm thick	m	0.10	0.80	2.21	0.12	3.13
75mm thick	m	0.12	0.96	2.88	0.48	4.32
100mm thick	m	0.14	1.12	3.64	0.60	5.36

	Unit	Hours	Hours £	Materials £	O & P £	Total £
Brick paving, 215 x 103 x 65mm laid to falls and cross falls bedding in cement mortar 15mm thick						
straight joints both ways						
bricks laid flat	m2	1.00	8.00	21.44	1.20	30.64
bricks laid on edge	m2	1.25	10.00	32.66	4.72	47.38
herringbone pattern						
bricks laid flat	m2	1.25	10.00	21.44	1.50	32.94
bricks laid on edge	m2	1.50	12.00	32.66	5.02	49.68
Preacast concrete paving flags spot bedded in cement mortar						
natural colour						
450 x 450 x 50mm	m2	0.50	4.00	11.22	0.60	15.82
600 x 600 x 50mm	m2	0.48	3.84	9.21	2.26	15.31
600 x 900 x 50mm	m2	0.45	3.60	8.55	1.92	14.07
coloured						
450 x 450 x 50mm	m2	0.50	4.00	12.43	0.60	17.03
600 x 600 x 50mm	m2	0.48	3.84	9.98	2.44	16.26
600 x 900 x 50mm	m2	0.15	1.20	9.15	1.68	12.03

	Unit	Specialist Price £	O & P £	Total £

ALTERATIONS AND REPAIRS

Shoring

	Unit	Specialist Price £	O & P £	Total £
Raking shoring consisting of two 175 x 175mm rakers, each 5m length, 50 x 150mm wall plate including sole pieces, steel dogs needles, cleats and bracing	nr	658.00	98.70	756.70
Raking shoring consisting of two 225 x 225mm rakers, each 8m length, 50 x 200mm wall plate including sole pieces, steel dogs needles, cleats, bracing and cutting through walls, floors and ceilings	nr	827.00	124.05	951.05
Dead shoring consisting of two 175 x 175mm legs including needle, sole plate, braces, wedges steel dogs and cutting through walls, floors and ceilings	nr	576.00	86.40	662.40
Flying shoring consisting of 100 x 150mm shore, 50 x 200 wall plates, 50 x 100mm struts and straining pieces, distance between walls being supported				
3m	nr	489.00	73.35	562.35
4m	nr	523.00	78.45	601.45
5m	nr	576.00	86.40	662.40
Strut existing window openings consisting of 125 x 50mm wall plates and struts	nr	123.00	18.45	141.45

	Unit	Hours	Hours £	Materials £	O & P £	Total £

Forming openings

Form opening for windows or
doors in existing walls in cement
mortar and make good to sides
of openings

	Unit	Hours	Hours £	Materials £	O & P £	Total £
75mm blockwork	m2	2.00	16.00	2.25	2.74	20.99
100mm blockwork	m2	2.22	17.76	2.25	3.00	23.01
140mm blockwork	m2	2.65	21.20	2.75	3.59	27.54
215mm blockwork	m2	2.80	22.40	3.30	3.86	29.56
half brick wall	m2	2.90	23.20	2.25	3.82	29.27
one brick wall	m2	3.78	30.24	2.88	4.97	38.09
one and a half brick wall	m2	4.80	38.40	3.57	6.30	48.27
two brick wall	m2	7.20	57.60	4.69	9.34	71.63

Form opening for windows or
doors in existing walls in lime
mortar and make good to sides
of openings

	Unit	Hours	Hours £	Materials £	O & P £	Total £
75mm blockwork	m2	1.80	14.40	2.25	2.50	19.15
100mm blockwork	m2	2.00	16.00	2.25	2.74	20.99
140mm blockwork	m2	2.40	19.20	2.75	3.29	25.24
215mm blockwork	m2	2.55	20.40	3.30	3.56	27.26
half brick wall	m2	2.60	20.80	2.25	3.46	26.51
one brick wall	m2	3.40	27.20	2.88	4.51	34.59
one and a half brick wall	m2	4.80	38.40	3.57	6.30	48.27
two brick wall	m2	6.50	52.00	4.69	8.50	65.19

Form opening for lintels above
openings in existing walls in cement
mortar and make good to sides
of openings

	Unit	Hours	Hours £	Materials £	O & P £	Total £
75mm blockwork	m2	2.98	23.84	1.20	3.76	28.80
100mm blockwork	m2	3.36	26.88	1.20	4.21	32.29
140mm blockwork	m2	3.87	30.96	1.60	4.88	37.44
215mm blockwork	m2	4.23	33.84	2.20	5.41	41.45
half brick wall	m2	6.30	50.40	1.60	7.80	59.80

	Unit	Hours	Hours £	Materials £	O & P £	Total £
one brick wall	m2	10.50	84.00	2.88	13.03	99.91
one and a half brick wall	m2	12.60	100.80	3.57	15.66	120.03
two brick wall	m2	16.98	135.84	4.69	21.08	161.61

Form opening for lintels above openings in existing walls in lime mortar and make good to sides of openings

75mm blockwork	m2	2.75	22.00	2.25	3.64	27.89
100mm blockwork	m2	3.00	24.00	2.25	3.94	30.19
140mm blockwork	m2	3.45	27.60	2.75	4.55	34.90
215mm blockwork	m2	3.86	30.88	3.30	5.13	39.31
half brick wall	m2	5.75	46.00	2.25	7.24	55.49
one brick wall	m2	9.45	75.60	2.88	11.77	90.25
one and a half brick wall	m2	11.40	91.20	3.57	14.22	108.99
two brick wall	m2	15.30	122.40	4.69	19.06	146.15

Form opening in reinforced concrete floor slab and make good to edges, slab thickness

100mm	m2	5.65	45.20	2.50	7.16	54.86
150mm	m2	7.86	62.88	2.75	9.84	75.47
200mm	m2	9.50	76.00	3.00	11.85	90.85
250mm	m2	11.57	92.56	3.25	14.37	110.18
300mm	m2	14.55	116.40	3.50	17.99	137.89

Form opening in reinforced concrete walls and make good to edges, wall thickness

100mm	m2	6.50	52.00	2.50	8.18	62.68
150mm	m2	7.20	57.60	2.75	9.05	69.40
200mm	m2	10.33	82.64	3.00	12.85	98.49
250mm	m2	12.10	96.80	3.25	15.01	115.06
300mm	m2	15.32	122.56	3.50	18.91	144.97

	Unit	Hours	Hours £	Materials £	O & P £	Total £

Filling openings

Take out existing door and frame
complete, make good to reveals
and piece up skirting to match
existing

	Unit	Hours	Hours £	Materials £	O & P £	Total £
single internal door	nr	6.00	48.00	9.88	8.68	66.56
single external door	nr	6.20	49.60	9.88	8.92	68.40
double internal door	nr	6.20	49.60	10.44	9.01	69.05
double external door	nr	6.40	51.20	10.44	9.25	70.89

Take out existing single door and
frame complete, fill in opening, fix
new skirting to match existing and
plaster both sides

	Unit	Hours	Hours £	Materials £	O & P £	Total £
100mm blockwork	m2	10.45	105.65	35.80	21.22	162.67
215mm blockwork	m2	12.47	126.07	52.32	26.76	205.15
half brick wall	m2	12.40	125.36	32.70	23.71	181.77
one brick wall	m2	14.60	147.61	58.97	30.99	237.56
one and a half brick wall	m2	16.37	165.50	81.25	37.01	283.76

Take out existing double doors and
frame complete, fill in opening, fix
new skirting to match existing and
plaster both sides

	Unit	Hours	Hours £	Materials £	O & P £	Total £
100mm blockwork	m2	14.59	147.50	42.20	28.46	218.16
215mm blockwork	m2	16.36	165.40	94.68	39.01	299.09
half brick wall	m2	15.54	157.11	54.37	31.72	243.20
one brick wall	m2	20.34	205.64	99.87	45.83	351.33
one and a half brick wall	m2	22.20	224.44	151.39	56.37	432.21

	Unit	Hours	Hours £	Materials £	O & P £	Total £

Underpinning

Excavate preliminary trench by
hand, maximum depth not
exceeding

	Unit	Hours	Hours £	Materials £	O & P £	Total £
1.00m	m3	3.70	29.60	0.00	4.44	34.04
1.50m	m3	4.15	33.20	0.00	4.98	38.18
2.00m	m3	4.40	35.20	0.00	5.28	40.48
2.50m	m3	6.10	48.80	0.00	7.32	56.12
3.00m	m3	7.25	58.00	0.00	8.70	66.70

Excavate trench by hand below
existing foundations, maximum
not exceeding

	Unit	Hours	Hours £	Materials £	O & P £	Total £
1.00m	m3	4.25	34.00	0.00	5.10	39.10
1.50m	m3	4.75	38.00	0.00	5.70	43.70
2.00m	m3	5.45	43.60	0.00	6.54	50.14
2.50m	m3	7.30	58.40	0.00	8.76	67.16
3.00m	m3	8.35	66.80	0.00	10.02	76.82

Excavate and backfill working
space, maximum depth not
exceeding

	Unit	Hours	Hours £	Materials £	O & P £	Total £
1.00m	m3	5.00	40.00	0.00	6.00	46.00
1.50m	m3	5.25	42.00	0.00	6.30	48.30
2.00m	m3	5.90	47.20	0.00	7.08	54.28
2.50m	m3	8.10	64.80	0.00	9.72	74.52
3.00m	m3	9.25	74.00	0.00	11.10	85.10

Open-boarded earthwork support
to sides of preliminary trenches,
distance between faces not
exceeding 2.0m

maximum depth not exceeding

	Unit	Hours	Hours £	Materials £	O & P £	Total £
1.0m	m2	0.30	2.40	1.55	0.59	4.54

	Unit	Hours	Hours £	Materials £	O & P £	Total £
maximum depth not exceeding 2.0m	m2	0.40	3.20	1.55	0.71	5.46
Close-boarded earthwork support to sides of preliminary trenches, distance between faces not exceeding 2.0m						
maximum depth not exceeding 1.0m	m2	0.85	6.80	2.64	1.42	10.86
maximum depth not exceeding 2.0m	m2	1.10	8.80	2.64	1.72	13.16
Open-boarded earthwork support to sides of excavation trenches, distance between faces not exceeding 2.0m						
maximum depth not exceeding 1.0m	m2	0.35	2.80	1.55	0.65	5.00
maximum depth not exceeding 2.0m	m2	0.45	3.60	1.55	0.77	5.92
Close-boarded earthwork support to sides of excavation trenches, distance between faces not exceeding 2.0m						
maximum depth not exceeding 1.0m	m2	0.95	7.60	2.64	1.54	11.78
maximum depth not exceeding 2.0m	m2	1.20	9.60	2.64	1.84	14.08

	Unit	Hours	Hours £	Materials £	O & P £	Total £
Cut away projecting concrete foundations, size						
150 x 150mm	m	0.45	3.60	0.00	0.54	4.14
150 x 225mm	m	0.55	4.40	0.00	0.66	5.06
150 x 300mm	m	0.65	5.20	0.00	0.78	5.98
Cut away projecting brickwork in footings, one brick thick						
one course	m	1.10	8.80	0.00	1.32	10.12
two courses	m	1.90	15.20	0.00	2.28	17.48
three courses	m	2.60	20.80	0.00	3.12	23.92
Prepare underside of existing foundations to receive the new work, width						
300mm	m	0.75	6.00	0.00	0.90	6.90
500mm	m	1.15	9.20	0.00	1.38	10.58
750mm	m	1.40	11.20	0.00	1.68	12.88
1000mm	m	1.60	12.80	0.00	1.92	14.72
1200mm	m	1.80	14.40	0.00	2.16	16.56
Load surplus excavated material into barrows, wheel and deposit in temporary spoil heaps, average distance						
15m	m3	1.25	10.00	0.00	1.50	11.50
25m	m3	1.45	11.60	0.00	1.74	13.34
50m	m3	1.70	13.60	0.00	2.04	15.64
Load surplus excavated material into barrows, wheel and deposit in temporary spoil heaps, average distance						
25m	m3	1.45	11.60	0.00	1.74	13.34
50m	m3	1.70	13.60	0.00	2.04	15.64

	Unit	Hours	Hours £	Materials £	O & P £	Total £
Load surplus excavated material into barrows, wheel and deposit in skips or lorries, average distance						
25m	m3	1.40	11.20	0.00	1.68	12.88
50m	m3	1.65	13.20	0.00	1.98	15.18
Level and compact bottom of excavation	m2	0.15	1.20	0.00	0.18	1.38
Site-mixed concrete 1:3:6 40mm aggregate in foundations to under-pinning, thickness						
150 to 300mm	m3	4.35	34.80	62.15	14.54	111.49
300 to 450mm	m3	4.05	32.40	62.15	14.18	108.73
over 450mm	m3	3.65	29.20	62.15	13.70	105.05
Site-mixed concrete 1:2:4 20mm aggregate in foundations to under-pinning, thickness						
150 to 300mm	m3	4.35	34.80	69.47	15.64	119.91
300 to 450mm	m3	4.05	32.40	69.47	15.28	117.15
over 450mm	m3	3.65	29.20	69.47	14.80	113.47
Plain vertical formwork to sides of underpinned foundations, height						
over 1m	m2	2.30	18.40	7.98	3.96	30.34
not exceeding 250mm	m	0.75	6.00	7.98	2.10	16.08
250 to 500mm	m	1.25	10.00	7.98	2.70	20.68
500mm to 1m	m	1.70	13.60	7.98	3.24	24.82

	Unit	Hours	Hours £	Materials £	O & P £	Total £
Plain vertical formwork to sides of underpinned foundations, left in, height						
over 1m	m2	2.15	17.20	19.40	5.49	42.09
not exceeding 250mm	m	0.65	5.20	6.90	1.82	13.92
250 to 500mm	m	1.15	9.20	12.76	3.29	25.25
500mm to 1m	m	1.60	12.80	19.40	4.83	37.03
High yield deformed steel reinforcement bars, straight or bent						
10mm diameter	m	0.04	0.32	0.26	0.09	0.67
12mm diameter	m	0.05	0.40	0.34	0.11	0.85
16mm diameter	m	0.06	0.48	0.54	0.15	1.17
20mm diameter	m	0.07	0.56	0.86	0.21	1.63
25mm diameter	m	0.08	0.64	1.40	0.31	2.35
Common bricks basic price £200 per thousand in cement mortar in underpinning						
one brick thick	m2	5.20	52.57	33.20	12.87	98.64
one and a half brick thick	m2	7.40	74.81	49.80	18.69	143.31
two brick thick	m2	9.35	94.53	86.40	27.14	208.07
Class A engineering bricks in basic price £350 per thousand in cement mortar in underpinning						
one brick thick	m2	5.40	54.59	57.26	16.78	128.63
one and a half brick thick	m2	7.60	76.84	85.89	24.41	187.13
two brick thick	m2	9.50	96.05	114.57	31.59	242.21
Hessian-based bitumen damp proof course, bedded in cement mortar, horizontal						
over 225mm wide	m2	0.35	3.54	8.14	1.75	13.43
112mm wide	m	0.05	0.51	0.95	0.22	1.67

	Unit	Hours	Hours £	Materials £	O & P £	Total £
Two courses of slates bedded in cement mortar, horizontal						
over 225mm wide	m2	2.90	29.32	27.88	8.58	65.78
225mm wide	m	0.85	8.59	5.68	2.14	16.41
Wedge and pin new work to soffit of existing with slates in cement mortar, width						
one brick thick	m	1.90	19.21	5.68	3.73	28.62
one and a half brick thick	m	2.40	24.26	7.91	4.83	37.00
two brick thick	m	2.80	28.31	11.67	6.00	45.97

Temporary screens

	Unit	Hours	Hours £	Materials £	O & P £	Total £
Erect, maintain and remove temporary screens consisting of 50 x 50mm softwood framing covered one side with						
hardboard (three uses)	m2	1.15	11.63	1.20	1.92	14.75
insulation board (three uses)	m2	1.15	11.63	1.65	1.99	15.27
polythene sheeting (three uses)	m2	0.80	8.09	0.12	1.23	9.44

SPOT ITEMS

Brickwork, blockwork and masonry

	Unit	Hours	Hours £	Materials £	O & P £	Total £
Take out existing fireplace, fill opening with 100mm thick block-work plastered one side, fix new skirting to match existing, make good flooring where hearth removed						
medium size fireplace	nr	12.00	96.00	17.55	17.03	130.58
large size fireplace	nr	14.00	112.00	21.69	20.05	153.74

	Unit	Hours	Hours £	Materials £	O & P £	Total £
Cut out existing projecting chimney breast, size 1350 x 350mm, including making good flooring and ceiling to match existing and plastering wall where breast removed						
one storey	nr	36.00	288.00	34.50	48.38	370.88
Take down chimney stack to 100mm below roof level, seal flue with slates in cement mortar, make good roof timbers						
slated roof	nr	46.00	368.00	87.55	68.33	523.88
tiled roof	nr	46.00	368.00	56.60	63.69	488.29
Cut out single brick in half brick wall and replace in gauged mortar						
commons	nr	0.25	2.53	0.20	0.41	3.14
facings	nr	0.30	3.03	0.40	0.51	3.95
Cut out decayed brickwork in half brick wall in areas 0.5 to 1m2 and replace in gauged mortar						
commons	nr	5.20	52.57	11.10	9.55	73.22
facings	nr	6.00	60.66	31.20	13.78	105.64
Cut out decayed brickwork in one brick wall in areas 0.5 to 1m2 and replace in gauged mortar						
commons	nr	7.60	76.84	22.20	14.86	113:89
facings	nr	8.80	88.97	38.80	19.17	146.93

	Unit	Hours	Hours £	Materials £	O & P £	Total £
Cut out vertical, horizontal or stepped cracks in half brick wall, replace average 350mm wide with new bricks in gauged mortar						
commons	m	2.75	27.80	4.15	4.79	36.75
facings	m	3.10	31.34	10.54	6.28	48.16
Cut out vertical, horizontal or stepped cracks in one brick wall, replace average 350mm wide with new bricks in gauged mortar						
commons	m	5.20	52.57	8.30	9.13	70.00
facings	m	5.80	58.64	21.04	11.95	91.63
Cut out defective brick-on-end soldier arch to half brick wall, and replace with new bricks in gauged mortar						
commons	m	3.00	30.33	6.30	5.49	42.12
facings	m	3.40	34.37	9.63	6.60	50.60
Cut out defective brick-on-end soldier arch to one brick wall, and replace with new bricks in gauged mortar						
commons	m	3.80	38.42	12.60	7.65	58.67
facings	m	4.20	42.46	19.26	9.26	70.98
Cut out defective terracotta air brick and replace						
215 x 65mm	nr	0.35	3.54	1.95	0.82	6.31
215 x 140mm	nr	0.50	5.06	2.62	1.15	8.83
215 x 215mm	nr	0.65	6.57	7.25	2.07	15.89

	Unit	Hours	Hours £	Materials £	O & P £	Total £
Rake out joints in gauged mortar, refix loose flashing and point up on completion						
horizontal	m	0.45	4.55	0.40	0.74	5.69
stepped	m	0.65	6.57	0.60	1.08	8.25
Rake out joints and point up in gauged mortar	m2	0.65	6.57	0.60	1.08	8.25
Cut out one course of half brick wall, insert hessian-based damp course 112mm wide and replace with new bricks in gauged mortar						
commons	m	2.00	20.22	2.54	3.41	26.17
facings	m	2.00	20.22	2.88	3.47	26.57
Cut out one course of one brick wall, insert hessian-based damp course 112mm wide and replace with new bricks in gauged mortar						
commons	m	3.25	32.86	5.08	5.69	43.63
facings	m	3.25	32.86	5.76	5.79	44.41
Cut out single block, replace with new in gauged mortar, thickness						
100mm	nr	0.30	3.03	1.10	0.62	4.75
140mm	nr	0.40	4.04	1.75	0.87	6.66
190mm	nr	0.50	5.06	2.20	1.09	8.34
255mm	nr	0.60	6.07	2.98	1.36	10.40
Rake out joints of random rubble walling and point up in gauged mortar						
flush pointing	m2	0.50	5.06	0.50	0.83	6.39
weather pointing	m2	0.55	5.56	0.50	0.91	6.97

	Unit	Hours	Hours £	Materials £	O & P £	Total £

Roofing

Take up roof coverings from pitched roof

	Unit	Hours	Hours £	Materials £	O & P £	Total £
tiles	m2	0.80	6.40	0.00	0.96	7.36
slates	m2	0.80	6.40	0.00	0.96	7.36
timber boarding	m2	1.00	8.00	0.00	1.20	9.20
metal sheeting	m2	0.20	1.60	0.00	0.24	1.84
flat sheeting	m2	0.30	2.40	0.00	0.36	2.76
corrugated sheeting	m2	0.30	2.40	0.00	0.36	2.76
underfelt	m2	0.10	0.80	0.00	0.12	0.92

Take up roof coverings from flat roof

	Unit	Hours	Hours £	Materials £	O & P £	Total £
bituminous felt	m2	0.25	2.00	0.00	0.30	2.30
metal sheeting	m2	0.30	2.40	0.00	0.36	2.76
woodwool slabs	m2	0.50	4.00	0.00	0.60	4.60
firrings	m2	0.20	1.60	0.00	0.24	1.84

Take up roof coverings from pitched roof, carefully lay aside for reuse

	Unit	Hours	Hours £	Materials £	O & P £	Total £
tiles	m2	1.10	8.80	0.00	1.32	10.12
slates	m2	1.10	8.80	0.00	1.32	10.12
metal sheeting	m2	0.50	4.00	0.00	0.60	4.60
flat sheeting	m2	0.60	4.80	0.00	0.72	5.52
corrugated sheeting	m2	0.60	4.80	0.00	0.72	5.52

Take up roof coverings from flat roof, carefully lay aside for reuse

	Unit	Hours	Hours £	Materials £	O & P £	Total £
metal sheeting	m2	0.60	4.80	0.00	0.72	5.52
woodwool slabs	m2	0.80	6.40	0.00	0.96	7.36

	Unit	Hours	Hours £	Materials £	O & P £	Total £
Inspect roof battens, refix loose and replace with new, size 38 x 25mm						
25% of area						
250mm centres	m2	0.14	1.42	0.30	0.26	1.97
450mm centres	m2	0.12	1.21	0.25	0.22	1.68
600mm centres	m2	0.10	1.01	0.20	0.18	1.39
50% of area						
250mm centres	m2	0.26	2.63	0.60	0.48	3.71
450mm centres	m2	0.16	1.62	0.50	0.32	2.44
600mm centres	m2	0.18	1.82	0.40	0.33	2.55
75% of area						
250mm centres	m2	0.36	3.64	0.90	0.68	5.22
450mm centres	m2	0.26	2.63	0.75	0.51	3.89
600mm centres	m2	0.22	2.22	0.60	0.42	3.25
100% of area						
250mm centres	m2	0.44	4.45	1.20	0.85	6.50
450mm centres	m2	0.32	3.24	1.00	0.64	4.87
600mm centres	m2	0.38	3.84	0.80	0.70	5.34
Remove single slipped slate and refix	nr	1.00	10.11	0.80	1.64	12.55
Remove single broken slate, renew with new Welsh blue slate						
405 x 255mm	nr	1.20	12.13	1.20	2.00	15.33
510 x 255mm	nr	1.20	12.13	2.30	2.16	16.60
610 x 305mm	nr	1.20	12.13	4.85	2.55	19.53
Remove slates in area approximately 1m2 and replace with Welsh blue slates previously laid aside	nr	1.60	16.18	0.00	2.43	18.60

	Unit	Hours	Hours £	Materials £	O & P £	Total £
Remove slates in area approximately 1m2 and replace with Welsh blue slates previously laid aside						
405 x 255mm	nr	1.80	18.20	47.22	9.81	75.23
510 x 255mm	nr	1.70	17.19	49.21	9.96	76.36
610 x 305mm	nr	1.60	16.18	52.39	10.28	78.85
Remove double course at eaves and fix new Welsh blue slates						
405 x 255mm	m	0.70	7.08	22.13	4.38	33.59
510 x 255mm	m	0.70	7.08	27.68	5.21	39.97
610 x 305mm	m	0.70	7.08	31.24	5.75	44.06
Remove single verge undercloak course and renew						
405 x 255mm	m	0.90	9.10	16.57	3.85	29.52
510 x 255mm	m	0.90	9.10	17.58	4.00	30.68
610 x 305mm	m	0.90	9.10	19.27	4.26	32.62
Remove single slipped tile and refix	nr	0.30	3.03	0.00	0.45	3.49
Remove single broken tile and renew						
Marley plain tile	nr	0.30	3.03	0.48	0.53	4.04
Marley Ludlow Plus tile	nr	0.30	3.03	0.72	0.56	4.32
Marley Modern tile	nr	0.30	3.03	1.10	0.62	4.75
Redland Renown tile	nr	0.30	3.03	1.10	0.62	4.75
Redland Norfolk tile	nr	0.30	3.03	0.95	0.60	4.58
Remove tiles in area approximately 1m2 and replace with tiles previously laid aside						
Marley plain tile	nr	1.80	18.20	24.56	6.41	49.17
Marley Ludlow Plus tile	nr	1.20	12.13	10.76	3.43	26.33
Marley Modern tile	nr	1.10	11.12	11.50	3.39	26.01

	Unit	Hours	Hours £	Materials £	O & P £	Total £
Redland Renown tile	nr	1.10	11.12	11.38	3.38	25.88
Redland Norfolk tile	nr	1.15	11.63	14.02	3.85	29.49
Take off defective ridge capping and refix including pointing in mortar	m	1.10	11.12	1.32	1.87	14.31

Carpentry and joinery

Take down, cut out or demolish structural timbers and load into skips

	Unit	Hours	Hours £	Materials £	O & P £	Total £
structural timbers						
50 x 100mm	m	0.10	0.80	0.00	0.12	0.92
50 x 150mm	m	0.12	0.96	0.00	0.14	1.10
75 x 100mm	m	0.14	1.12	0.00	0.17	1.29
75 x 150mm	m	0.16	1.28	0.00	0.19	1.47
100 x 150mm	m	0.18	1.44	0.00	0.22	1.66
100 x 200mm	m	0.20	1.60	0.00	0.24	1.84
roof boarding	m2	0.28	2.24	0.00	0.34	2.58
floor boarding	m2	0.22	1.76	0.00	0.26	2.02
stud partition plasterboard both sides	m2	0.75	6.00	0.00	0.90	6.90
skirtings and grounds						
100mm high	m	0.08	0.64	0.00	0.10	0.74
150mm high	m	0.09	0.72	0.00	0.11	0.83
200mm high	m	0.10	0.80	0.00	0.12	0.92
rails						
50mm high	m	0.05	0.40	0.00	0.06	0.46
75mm high	m	0.06	0.48	0.00	0.07	0.55
100mm high	m	0.07	0.56	0.00	0.08	0.64
fittings						
wall cupboards	nr	0.25	2.00	0.00	0.30	2.30
floor units	nr	0.20	1.60	0.00	0.24	1.84
sink units	nr	0.25	2.00	0.00	0.30	2.30
staircase, 900mm wide						
straight flight	nr	4.00	32.00	0.00	4.80	36.80
landing	nr	1.50	12.00	0.00	1.80	13.80

	Unit	Hours	Hours £	Materials £	O & P £	Total £
doors, frames and linings						
single, internal	nr	0.40	3.20	0.00	0.48	3.68
single, external	nr	0.60	4.80	0.00	0.72	5.52
double, internal	nr	0.60	4.80	0.00	0.72	5.52
double, external	nr	0.80	6.40	0.00	0.96	7.36
windows						
casement, 1200 x 900mm	nr	0.50	4.00	0.00	0.60	4.60
casement, 1800 x 900mm	nr	0.60	4.80	0.00	0.72	5.52
sash, 900 x 1500mm	nr	0.80	6.40	0.00	0.96	7.36
sash, 1800 x 900mm	nr	0.90	7.20	0.00	1.08	8.28
ironmongery						
bolt	nr	0.20	1.60	0.00	0.24	1.84
deadlock	nr	0.25	2.00	0.00	0.30	2.30
mortice lock	nr	0.35	2.80	0.00	0.42	3.22
mortice latch	nr	0.35	2.80	0.00	0.42	3.22
cylinder lock	nr	0.25	2.00	0.00	0.30	2.30
door closer	nr	0.35	2.80	0.00	0.42	3.22
casement stay	nr	0.15	1.20	0.00	0.18	1.38
casement fastener	nr	0.15	1.20	0.00	0.18	1.38
toilet roll holder	nr	0.15	1.20	0.00	0.18	1.38
shelf bracket	nr	0.15	1.20	0.00	0.18	1.38
Cut out defective joists or rafters and replace with new						
50 x 75mm	nr	0.35	3.54	1.16	0.70	5.40
50 x 100mm	nr	0.40	4.04	1.20	0.79	6.03
50 x 150mm	nr	0.60	6.07	1.80	1.18	9.05
75 x 100mm	nr	0.65	6.57	2.42	1.35	10.34
75 x 150mm	nr	0.75	7.58	2.94	1.58	12.10
Cut out defective skirting and renew						
75mm high	m	0.30	3.03	1.90	0.74	5.67
100mm high	m	0.35	3.54	2.22	0.86	6.62
150mm high	m	0.40	4.04	2.87	1.04	7.95

	Unit	Hours	Hours £	Materials £	O & P £	Total £
Ease, adjust and oil						
door	nr	0.60	6.07	0.00	0.91	6.98
casement window	nr	0.40	4.04	0.00	0.61	4.65
sash window including renewing cords	nr	1.50	15.17	4.22	2.91	22.29
Take up defective flooring and replace with 25mm thick plain edged boarding						
areas less than 1m2	m2	1.50	15.17	11.50	4.00	30.66
areas more than 1m2	m2	1.20	12.13	11.50	3.54	27.18
Refix loose floorboards including punching in protruding nails	m2	0.20	2.02	0.00	0.30	2.33
Take down existing door, lay aside, piece up frame or lining where butts removed, rehang door to opposite hand on existing butts and hardware						
single, internal	nr	1.50	15.17	1.50	2.50	19.16
single, external	nr	1.70	17.19	1.50	2.80	21.49

Finishings

Take down plasterboard sheeting from

	Unit	Hours	Hours £	Materials £	O & P £	Total £
studded walls	m2	0.35	2.80	0.00	0.42	3.22
ceilings	m2	0.40	3.20	0.00	0.48	3.68

Hack off plaster from

	Unit	Hours	Hours £	Materials £	O & P £	Total £
walls	m2	0.25	2.00	0.00	0.30	2.30
ceiling	m2	0.30	2.40	0.00	0.36	2.76

	Unit	Hours	Hours £	Materials £	O & P £	Total £
Make good plaster to walls where wall removed						
100mm wide	m2	0.95	9.60	1.50	1.67	12.77
150mm wide	m2	1.05	10.62	1.60	1.83	14.05
200mm wide	m2	1.15	11.63	1.70	2.00	15.33
Make good plaster to ceilings where wall removed						
100mm wide	m2	1.10	11.12	1.50	1.89	14.51
150mm wide	m2	1.15	11.63	1.60	1.98	15.21
200mm wide	m2	1.20	12.13	1.70	2.07	15.91
Cut out cracks in plasterwork and make good						
walls	m2	0.35	3.54	0.25	0.57	4.36
ceiling	m2	0.40	4.04	0.25	0.64	4.94
Hack off wall tiling and make good surface	m2	0.75	7.58	4.56	1.82	13.96

Plumbing

	Unit	Hours	Hours £	Materials £	O & P £	Total £
Remove sanitary fittings and supports, seal off supply and waste pipes and prepare to receive new fittings						
bath	nr	3.50	28.00	10.00	5.70	43.70
sink	nr	3.50	28.00	10.00	5.70	43.70
lavatory basin	nr	3.25	26.00	6.00	4.80	36.80
WC	nr	3.75	30.00	6.00	5.40	41.40
shower cubicle	nr	4.00	32.00	8.00	6.00	46.00
bidet	nr	3.25	26.00	6.00	4.80	36.80

	Unit	Hours	Hours £	Materials £	O & P £	Total £
Take down length of existing gutter, prepare ends and install new length of gutter to existing brackets						
PVC-U						
76mm	nr	0.60	6.07	7.10	1.97	15.14
112mm	nr	0.65	6.57	8.70	2.29	17.56
cast iron						
100mm	nr	1.12	11.32	21.52	4.93	37.77
Take down existing brackets from fascias and replace with galvanised steel repair brackets at 1m centres	m	0.30	3.03	2.15	0.78	5.96
Take down length of existing pipe, prepare ends and install new length of pipe						
PVC-U						
68mm diameter	nr	0.75	7.58	7.87	2.32	17.77
68mm square	nr	0.75	7.58	8.64	2.43	18.66
cast iron						
75mm	nr	0.90	9.10	31.62	6.11	46.83
100mm	nr	1.05	10.62	43.88	8.17	62.67
Take down fittings and install new						
PVC-U						
68mm diameter bend	nr	0.50	5.06	2.94	1.20	9.19
68mm offset	nr	0.50	5.06	6.85	1.79	13.69
68mm branch	nr	0.50	5.06	7.00	1.81	13.86
68mm shoe	nr	0.75	7.58	6.33	2.09	16.00
Cast iron						
75mm diameter bend	nr	0.60	6.07	5.90	1.79	13.76
75mm offset	nr	0.60	6.07	14.88	3.14	24.09
75mm branch	nr	0.60	6.07	6.47	1.88	14.42
75mm shoe	nr	0.60	6.07	13.96	3.00	23.03

	Unit	Hours	Hours £	Materials £	O & P £	Total £
Cast iron						
100mm diameter bend	nr	0.70	7.08	19.82	4.03	30.93
100mm offset	nr	0.70	7.08	27.44	5.18	39.69
100mm branch	nr	0.70	7.08	7.92	2.25	17.25
100mm shoe	nr	0.70	7.08	16.70	3.57	27.34
Cut out 500mm length of copper pipe, install new pipe with compression fittings at each end						
15mm	nr	0.80	8.09	3.88	1.80	13.76
22mm	nr	0.90	9.10	6.80	2.38	18.28
Take off existing radiator valve and replace with new	nr	0.90	9.10	8.86	2.69	20.65
Take out existing galvanised steel water storage tank and install new plastic tank complete with ball valve, lid and insulation including cutting holes, make up pipework and connectors to existing pipework						
PC4, 16 litres	nr	3.60	36.40	14.24	7.60	58.23
PC15, 68 litres	nr	3.70	37.41	36.64	11.11	85.15
PC20, 91 litres	nr	3.75	37.91	38.35	11.44	87.70
PC25, 114 litres	nr	3.80	38.42	47.96	12.96	99.33
PC40, 182 litres	nr	4.90	49.54	76.80	18.95	145.29
PC50, 227 litres	nr	5.00	50.55	81.89	19.87	152.31

Glazing

	Unit	Hours	Hours £	Materials £	O & P £	Total £
Hack out glass and remove	m2	0.45	4.55	0.00	0.68	5.23
Clean rebates, remove sprigs or clips and prepare for reglazing	m	0.20	2.02	0.00	0.30	2.33

	Unit	Hours	Hours £	Materials £	O & P £	Total £

Painting

Prepare, wash down painted
surfaces, rub down to receive new
paintwork

brickwork	m2	0.14	1.42	0.00	0.21	1.63
blockwork	m2	0.15	1.52	0.00	0.23	1.74
plasterwork	m2	0.12	1.21	0.00	0.18	1.40

Prepare, wash down previously
painted wood surfaces, rub down
to receive new paintwork surfaces

over 300mm girth	m2	0.28	2.83	0.00	0.42	3.26
isolated surfaces not exceeding 300mm girth	m	0.10	1.01	0.00	0.15	1.16
isolated surfaces not exceeding 0.5m2	nr	0.12	1.21	0.00	0.18	1.40

Wallpapering

Strip off one layer of existing
paper, stop cracks and rub down
to receive new paper

woodchip						
walls	m2	0.18	1.82	0.00	0.27	2.09
walls in staircase areas	m2	0.20	2.02	0.00	0.30	2.33
ceilings	m2	0.22	2.22	0.00	0.33	2.56
ceilings in staircase areas	m2	0.24	2.43	0.00	0.36	2.79
vinyl						
walls	m2	0.23	2.33	0.00	0.35	2.67
walls in staircase areas	m2	0.25	2.53	0.00	0.38	2.91
ceilings	m2	0.27	2.73	0.00	0.41	3.14
ceilings in staircase areas	m2	0.29	2.93	0.00	0.44	3.37

	Unit	Hours	Hours £	Materials £	O & P £	Total £
standard patterned						
walls	m2	0.21	2.12	0.00	0.32	2.44
walls in staircase areas	m2	0.23	2.33	0.00	0.35	2.67
ceilings	m2	0.25	2.53	0.00	0.38	2.91
ceilings in staircase areas	m2	0.27	2.73	0.00	0.41	3.14

Part Two

APPROXIMATE ESTIMATING

	Unit	Total £

APPROXIMATE ESTIMATING

The following rates are based upon
the unit rates in Part One of this
book but have been rounded off for
ease of use

Excavation and filling

Excavate for trench including
supporting sides, level and ram
bottom, part return, fill and ram,
part load into skip, width 600mm

	Unit	Total £
by hand, depth		
0.75m	m	22.00
1.00m	m	27.00
1.50m	m	38.00
2.00m	m	48.00
by machine, depth		
0.75m	m	15.00
1.00m	m	18.00
1.50m	m	24.00
2.00m	m	30.00

Filling in layers by hand over
250mm thick

	Unit	Total £
surplus excavated material	m3	15.00
sand	m3	35.00
hardcore	m3	30.00

Filling in layers by machine over
250mm thick

	Unit	Total £
surplus excavated material	m3	8.00
sand	m3	30.00
hardcore	m3	25.00

	Unit	Total £
Concrete work		
Ready mixed concrete 1:3:6 40mm aggregate in foundations, size		
600 x 225mm	m	15.00
750 x 225mm	m	17.50
Ready mixed concrete 1:2:4 20mm aggregate, in beds including mesh reinforcement and trowelling smooth		
beds, thickness		
100mm	m2	18.00
150mm	m2	24.00
200mm	m2	30.00
Ready mixed concrete 1:2:4 20mm aggregate, in slabs including mesh reinforcement and formwork to soffit		
100mm	m2	68.00
150mm	m2	74.00
200mm	m2	80.00
Ready mixed concrete 1:2:4 20mm aggregate, in walls including mesh reinforcement and formwork both sides		
100mm	m2	74.00
150mm	m2	81.00
200mm	m2	88.00

	Unit	Total £

Brickwork and blockwork

Cavity wall in cement mortar
in foundations including
forming cavity and wall ties
between outer leaf in common
bricks £140 per 1000 and
inner leaf of

	Unit	Total £
common brick £140 per 1000	m2	89.00
blockwork 100mm thick	m2	77.00

Cavity wall in cement mortar
in foundations including
forming cavity and wall ties
between outer leaf in common
bricks £200 per 1000 and
inner leaf of

	Unit	Total £
common brick £200 per 1000	m2	99.00
blockwork 100mm thick	m2	82.00

Cavity wall in cement mortar
in foundations including
forming cavity and wall ties
between outer leaf in engineering
bricks £250 per 1000 and
inner leaf of

	Unit	Total £
engineering bricks £250 per 1000	m2	110.00
blockwork 100mm thick	m2	88.00

Cavity wall in cement mortar
in foundations including forming
cavity and wall ties between outer
leaf in engineering bricks £350 per
1000 and inner leaf of

	Unit	Total £
engineering bricks £350 per 1000	m2	129.00
blockwork 100mm thick	m2	97.00

	Unit	Total £
Cavity wall in gauged mortar including forming cavity and wall ties between outer leaf in facing bricks £250 per 1000 and inner leaf of		
common brick £140 per 1000	m2	67.00
common brick £200 per 1000	m2	112.00
blockwork 100mm thick	m2	95.00
Cavity wall in gauged mortar including forming cavity and wall ties between outer leaf in facing bricks £400 per 1000 and inner leaf of		
common brick £140 per 1000	m2	117.00
common brick £200 per 1000	m2	122.00
blockwork 100mm thick	m2	105.00
Cavity wall in gauged mortar including forming cavity and wall ties between outer leaf in facing bricks £500 per 1000 and inner leaf of		
common brick £140 per 1000	m2	123.00
common brick £200 per 1000	m2	128.00
blockwork 100mm thick	m2	111.00

Masonry

	Unit	Total £
Cavity wall in gauged mortar including forming cavity and wall ties between outer leaf in irregular coursed rubble walling 150mm wide and inner leaf of		
common brick £140 per 1000	m2	117.00
common brick £200 per 1000	m2	122.00

	Unit	Total £
Cavity wall in gauged mortar including forming cavity and wall ties between outer leaf in coursed rubble walling 150mm wide and inner leaf of		
common brick £140 per 1000	m2	122.00
common brick £200 per 1000	m2	127.00

Carpentry and joinery

19mm thick butt-jointed flooring fixed to softwood joists size

	Unit	Total
50 x 100mm	m2	22.00
50 x 125mm	m2	23.00
50 x 150mm	m2	24.00

25mm thick butt-jointed flooring fixed to softwood joists size

	Unit	Total
50 x 100mm	m2	25.00
50 x 125mm	m2	26.00
50 x 150mm	m2	27.00

19mm thick tongued and grooved flooring fixed to softwood joists size

	Unit	Total
50 x 100mm	m2	25.00
50 x 125mm	m2	26.00
50 x 150mm	m2	27.00

25mm thick tongued and grooved flooring fixed to softwood joists size

	Unit	Total
50 x 100mm	m2	27.00
50 x 125mm	m2	28.00
50 x 150mm	m2	29.00

	Unit	Total £
18mm thick chipboard butt-jointed flooring fixed to softwood joists size		
50 x 100mm	m2	17.00
50 x 125mm	m2	18.00
50 x 150mm	m2	19.00
25mm thick chipboard butt-jointed flooring fixed to softwood joists size		
50 x 100mm	m2	19.00
50 x 125mm	m2	20.00
50 x 150mm	m2	21.00
25mm thick chipboard tongued and grooved flooring fixed to softwood joists size		
50 x 100mm	m2	21.00
50 x 125mm	m2	22.00
50 x 150mm	m2	23.00
Standard flush hardboard-faced internal door including lining and stops, architraves and fixing only hardware		
35mm thick		
610 x 1981mm	nr	124.00
686 x 1981mm	nr	126.00
762 x 1981mm	nr	128.00
40mm thick		
610 x 1981mm	nr	127.00
686 x 1981mm	nr	129.00
762 x 1981mm	nr	131.00

	Unit	Total £
Standard flush sapele-faced internal door including lining and stops, architraves and fixing only hardware		
35mm thick		
610 x 1981mm	nr	139.00
686 x 1981mm	nr	141.00
762 x 1981mm	nr	143.00
40mm thick		
610 x 1981mm	nr	144.00
686 x 1981mm	nr	146.00
762 x 1981mm	nr	148.00
Standard flush hardboard-faced half-hour fire-check internal door including lining and stops, architraves and fixing only hardware		
44mm thick		
686 x 1981mm	nr	175.00
762 x 1981mm	nr	178.00
726 x 2040mm	nr	182.00
826 x 2040mm	nr	186.00
Standard flush sapele-faced half-hour fire-check internal door including lining and stops, architraves and fixing only hardware		
44mm thick		
686 x 1981mm	nr	200.00
762 x 1981mm	nr	205.00
726 x 2040mm	nr	210.00
826 x 2040mm	nr	215.00

	Unit	Total £
Standard flush hardboard-faced external door including frame, architraves and fixing only hardware		
44mm thick		
762 x 1981mm	nr	210.00
838 x 1981mm	nr	220.00

Finishings

One coat 'Universal' plaster to walls, plasterboard and skim to ceilings in rooms, storey height 2.4m

floor area		
9m2	nr	335.00
12m2	nr	415.00
15m2	nr	495.00
18m2	nr	555.00
21m2	nr	620.00
24m2	nr	700.00
27m2	nr	760.00
30m2	nr	820.00

Plumbing and heating

PVC-U rainwater goods

one-storey gable end		
terraced	nr	200.00
semi-detached	nr	240.00
detached	nr	280.00
two-storey gable end		
terraced	nr	250.00
semi-detached	nr	285.00
detached	nr	310.00

	Unit	Total £
three-storey gable end		
terraced	nr	315.00
semi-detached	nr	335.00
detached	nr	370.00
one-storey hipped end		
terraced	nr	200.00
semi-detached	nr	300.00
detached	nr	380.00
two-storey hipped end		
terraced	nr	280.00
semi-detached	nr	360.00
detached	nr	440.00
three-storey hipped end		
terraced	nr	330.00
semi-detached	nr	420.00
detached	nr	510.00
Cast iron rainwater goods		
one-storey gable end		
terraced	nr	530.00
semi-detached	nr	590.00
detached	nr	625.00
two-storey gable end		
terraced	nr	690.00
semi-detached	nr	740.00
detached	nr	780.00
three-storey gable end		
terraced	nr	810.00
semi-detached	nr	930.00
detached	nr	970.00

	Unit	Total £
one-storey hipped end		
terraced	nr	510.00
semi-detached	nr	680.00
detached	nr	880.00
two-storey hipped end		
terraced	nr	660.00
semi-detached	nr	840.00
detached	nr	1060.00
three-storey hipped end		
terraced	nr	780.00
semi-detached	nr	1010.00
detached	nr	1200.00

Central heating

one-storey house overall size		
(5 radiators)	nr	1,800.00
(6 radiators)	nr	1,950.00
(7 radiators)	nr	2,075.00
two-storey house overall size		
(8 radiators)	nr	2,450.00
(9 radiators)	nr	2,650.00
(10 radiators)	nr	2,950.00
three-storey house overall size		
(12 radiators)	nr	4,100.00
(14 radiators)	nr	4,350.00
(15 radiators)	nr	4,500.00

Painting

One mist coat and two coats of
emulsion paint to walls and ceilings
in rooms, storey height 2.4m

floor area		
9m2	nr	220.00

	Unit	Total £
12m2	nr	260.00
15m2	nr	305.00
18m2	nr	335.00
21m2	nr	365.00
24m2	nr	410.00
27m2	nr	435.00
30m2	nr	470.00

Wallpapering

Prepare and hang wallpaper £5 per
roll to walls in rooms, storey
height 2.4m

floor area

9m2	nr	140.00
12m2	nr	165.00
15m2	nr	190.00
18m2	nr	200.00
21m2	nr	215.00
24m2	nr	235.00
27m2	nr	250.00
30m2	nr	260.00

Prepare and hang wallpaper £7 per
roll to walls in rooms, storey
height 2.4m

floor area

9m2	nr	155.00
12m2	nr	180.00
15m2	nr	210.00
18m2	nr	220.00
21m2	nr	235.00
24m2	nr	260.00
27m2	nr	275.00
30m2	nr	285.00

	Unit	Total £

Prepare and hang wallpaper £9 per
roll to walls in rooms, storey
height 2.4m

floor area		
9m2	nr	170.00
12m2	nr	200.00
15m2	nr	225.00
18m2	nr	240.00
21m2	nr	255.00
24m2	nr	285.00
27m2	nr	300.00
30m2	nr	315.00

Drainage

Excavate trench by hand, lay
100mm diameter 'Hepsleve' pipe,
granular bed and surround, trench
depth

0.50m	m	30.00
0.75m	m	37.00
1.00m	m	42.00
1.25m	m	55.00
1.50m	m	65.00
1.75m	m	74.00
2.00m	m	85.00
2.25m	m	94.00
2.50m	m	106.00
2.75m	m	115.00
3.00m	m	129.00

Excavate trench by hand, lay
150mm diameter 'Hepsleve' pipe,
granular bed and surround, trench
depth

0.50m	m	41.00
0.75m	m	48.00

	Unit	Total £
1.00m	m	53.00
1.25m	m	66.00
1.50m	m	76.00
1.75m	m	85.00
2.00m	m	96.00
2.25m	m	105.00
2.50m	m	117.00
2.75m	m	126.00
3.00m	m	140.00

Excavate trench by machine, lay
100mm diameter 'Hepsleve' pipe,
granular bed and surround, trench
depth

	Unit	Total £
0.50m	m	13.00
0.75m	m	26.00
1.00m	m	31.00
1.25m	m	39.00
1.50m	m	42.00
1.75m	m	47.00
2.00m	m	53.00
2.25m	m	61.00
2.50m	m	68.00
2.75m	m	75.00
3.00m	m	81.00

Excavate trench by machine, lay
100mm diameter 'Hepsleve' pipe,
granular bed and surround, trench
depth

	Unit	Total £
0.50m	m	34.00
0.75m	m	37.00
1.00m	m	42.00
1.25m	m	50.00
1.50m	m	53.00
1.75m	m	58.00
2.00m	m	64.00
2.25m	m	72.00

	Unit	Total £
2.50m	m	79.00
2.75m	m	86.00
3.00m	m	92.00

Manhole complete including hand
excavation, concrete base and
benching, engineering brickwork,
channels and cast iron cover,
depth

1.00m	nr	415.00
1.50m	nr	560.00
2.00m	nr	725.00

Manhole complete including machine
excavation, concrete base and
benching, engineering brickwork,
channels and cast iron cover,
depth

1.00m	nr	400.00
1.50m	nr	515.00
2.00m	nr	630.00

Part Three

PLANT AND TOOL HIRE

	24 hours	Additional 24 hours	Week
	£	£	£

TOOL AND EQUIPMENT HIRE

The selected hire rates are based on those charged by Jewson Hire Point who have over 190 branches in the UK. Telephone 0800 539766 for details of your local branch and to check the prices currently available. These prices exclude VAT.

CONCRETE AND CUTTING EQUIPMENT

Concrete mixers

	24 hours	Additional 24 hours	Week
Petrol, 4/3 with stand	12.00	4.80	24.00
Electric, 4/3 with stand	10.00	4.00	20.00
Forced action	42.00	14.00	70.00
Diesel, 5/3.5	15.60	5.20	26.00
Mortar mixers			
diesel	78.00	26.00	130.00

Vibrating pokers

	24 hours	Additional 24 hours	Week
Pokers			
petrol	25.00	10.00	50.00
electric	24.20	8.80	44.00
electric, lightweight	20.90	7.60	38.00
extra poker shaft, 25/36/60mm	11.00	4.00	20.00
high frequency	32.56	13.00	65.00

Power floats

	24 hours	Additional 24 hours	Week
Floats			
600mm	33.00	12.00	60.00
900mm	33.00	12.00	60.00

	24 hours	Additional 24 hours	Week
	£	£	£
Concrete floats	13.20	4.80	24.00

Vibrating screeds

Screed units			
with 5m beam	38.50	14.00	70.00
with 3.25-5.20m beam	44.00	16.00	80.00
roller screed unit	99.00	36.00	180.00

Floor preparation units

Floor plane/roof de-chipper			
petrol	44.00	16.00	80.00
electric	44.00	16.00	80.00
Scrabbler			
single head	15.00	5.00	25.00
triple head	15.00	5.00	25.00
Electric needle gun	27.00	9.00	45.00
Diesel floor grinder	44.00	16.00	80.00
Diamond concrete planer	44.00	16.00	80.00
Needle chipping gun	15.00	5.00	25.00

Floor saws

Petrol floor saws			
350mm	44.00	16.00	80.00
450mm	44.00	16.00	80.00

Disc cutters

Cutters			
electric, 300mm	18.00	7.20	36.00
two stroke, 300mm	19.00	7.60	38.00
two stroke, 350mm	25.00	10.00	50.00
electric wall chasers	30.00	12.00	60.00
brick/wall cutter	50.00	20.00	100.00

	24 hours	Additional 24 hours	Week
	£	£	£

Block and slab splitters

Splitters
hydraulic	20.00	8.00	40.00
manual	17.00	6.80	34.00

Saw benches

Masonry saw bench
petrol, 350mm	44.00	16.00	80.00
petrol, 400mm	44.00	16.00	80.00
electric, 350mm	44.00	16.00	80.00

EXCAVATING AND SITE PLANT

Excavators

Mini excavator, diesel, rubber tracks	100.00	56.00	280.00

Air compressors

Compressor with breaker/hose
single tool	45.00	18.00	90.00
twin tool	60.00	24.00	120.00

Loaders and carriers

Tracked power barrow	60.00	24.00	120.00
Tracked skip loader	60.00	24.00	120.00
Powered wheel barrow	30.00	10.00	50.00
Wheeled skip loader	60.00	24.00	120.00

Dumpers

Mini dumper high tip, 750kg capacity	100.00	40.00	200.00
Swivel dumper, 2 ton	100.00	40.00	200.00

	24 hours	Additional 24 hours	Week
	£	£	£

ACCESS AND SITE EQUIPMENT

Ladders

Multi-purpose ladder	11.50	4.60	23.00
Double ladder, alloy			
14 feet	10.00	4.00	20.00
25 feet	13.00	5.20	26.00
31 feet	15.00	6.00	30.00
Triple ladder, alloy			
30 feet	12.00	4.80	24.00
35 feet, rope operated	26.40	9.60	48.00
41 feet, rope operated	26.40	9.60	48.00
47 feet, rope operated	33.00	12.00	60.00
Roof ladder			
14 feet	10.00	4.00	20.00
17 feet	10.00	4.00	20.00

Props

Shoring props			
type 0	0.00	0.50	2.50
type 1	0.00	0.00	2.50
type 2	0.00	0.50	2.50
type 3	0.00	0.80	4.00
type 4	0.00	0.80	4.00

Chimney scaffold

Scaffold			
single unit	42.00	14.00	70.00
twin unit	48.00	16.00	80.00
four unit	84.00	28.00	140.00

	24 hours	Additional 24 hours	Week
	£	£	£

Rubbish chutes

Chutes
1m section	3.60	1.20	6.00
funnel	3.60	1.20	6.00
frame	3.60	1.20	6.00

Trestles

Adjustable trestle
size 1	0.00	0.50	2.50
size 2	0.00	0.50	2.50
size 3	0.00	0.50	2.50

Painters' trestle
6 feet	5.00	2.00	10.00
8 feet	6.00	2.40	12.00

Staging boards

Boards size
10 feet/3.0m	6.00	2.40	12.00
12 feet/3.6m	7.00	2.80	14.00
14 feet/4.2m	8.00	3.20	16.00
16 feet/4.8m	10.00	4.00	20.00
18 feet/5.4m	11.00	4.40	22.00
22 feet/6.0m	13.00	5.20	26.00

Alloy towers

Single width, height
2.34m	22.80	7.60	38.00
3.27m	28.20	9.40	47.00
3.73m	32.40	10.80	54.00
4.66m	34.80	11.60	58.00
5.59m	38.40	12.80	64.00
6.52m	42.00	14.00	70.00
7.45m	51.60	17.20	86.00

	24 hours	Additional 24 hours	Week
	£	£	£
7.91m	53.40	17.80	89.00
8.84m	60.00	20.00	100.00
9.30m	66.00	22.00	110.00
Double width, height			
2.34m	34.80	11.60	58.00
3.27m	37.20	12.40	62.00
3.73m	39.00	13.00	65.00
4.66m	43.20	14.40	72.00
5.59m	48.00	16.00	80.00
6.52m	51.00	17.00	85.00
7.45m	57.00	19.00	95.00
7.91m	59.40	19.80	99.00
8.84m	68.40	22.80	114.00
10.23m	76.80	25.60	128.00
11.16m	85.80	28.60	143.00
12.55m	99.60	33.20	166.00
13.94m	108.60	36.20	181.00
14.87m	114.00	38.00	190.00
15.80m	119.40	39.80	199.00

MATERIAL HANDLING

Lifting and moving

	24 hours	Additional 24 hours	Week
Lifters			
Pavac slab lifter	20.00	8.00	40.00
Pavac slab lifter with hand pad			
and hose	5.00	2.00	10.00
Wheelbarrow	4.00	1.60	8.00
Engine crane	18.00	7.20	36.00
Plaster board jack lift	12.50	5.00	25.00

COMPACTION

Plate compactors

	24 hours	Additional 24 hours	Week
Compactors			
16in	25.00	10.00	50.00
18in	23.00	9.20	46.00
20in	23.00	9.20	46.00

	24 hours	**Additional 24 hours**	**Week**
	£	**£**	**£**

Rammers

Tamper rammers

9in, petrol	27.50	10.00	50.00
11in, petrol	27.50	10.00	50.00

Vibrating rollers

Single drum vibrating roller

BW55E	48.00	16.00	80.00
BW75E/71E	44.00	16.00	80.00

BREAKING AND DEMOLITION

Breakers

Demolition hammers

medium duty, electric	15.00	6.00	30.00
heavy duty, electric	18.00	7.20	36.00
Road breakers			
electric	30.00	12.00	60.00
2 stroke	45.00	18.00	90.00

Hydraulic breakers

Breakers

light duty, petrol	37.50	15.00	75.00
heavy duty, petrol	45.00	18.00	90.00
diesel	50.00	20.00	100.00

POWER TOOLS

Drills

Cordless drill/screwdriver	12.00	4.80	24.00
1/2in chuck cordless impact	15.00	6.00	30.00
1/2in chuck 2 speed impact	7.00	2.80	14.00
24mm rotary hammer	12.00	4.80	24.00
30mm rotary hammer	30.00	5.20	26.00

	24 hours	Additional 24 hours	Week
	£	£	£
35mm rotary hammer	17.00	6.80	34.00
Angle head drill	13.00	5.20	26.00
Cordless hammer	16.00	6.40	32.00

Impact wrenches

	24 hours	Additional 24 hours	Week
Wrenches			
electric	16.50	6.00	30.00
2 stroke	71.50	26.00	130.00
electric, heavy duty	33.00	12.00	60.00

Grinders

	24 hours	Additional 24 hours	Week
Mini grinder, 4.5in	7.50	3.00	15.00
Angle grinder/cutter, 9in	10.00	4.00	20.00

Saws

	24 hours	Additional 24 hours	Week
Reciprocating saw			
cordless, 18v	14.00	5.60	28.00
cordless, 24v	15.00	6.00	30.00
Circular saw			
9.25in	13.00	5.20	26.00
cordless, 6.50in, 24v	15.00	6.00	30.00
Door trimmer	27.50	11.00	55.00
Chop/cut cut off, 14in	27.50	10.00	50.00

Woodworking

	24 hours	Additional 24 hours	Week
Plane, 3.25in	13.00	5.20	26.00
Router	13.00	5.20	26.00
Laminate trimmer	13.00	5.20	26.00
Worktop jig	15.00	6.00	30.00
Jig saw	10.00	4.00	20.00

Fixing equipment

	24 hours	Additional 24 hours	Week
Cordless nailing gun	22.00	8.00	40.00
Air nail gun with compressor	50.00	20.00	100.00
Cartridge hammer	16.50	6.00	30.00

	24 hours	Additional 24 hours	Week
	£	£	£
Hammer tacker	5.00	2.00	10.00
Autofeed electric screwdriver	18.00	7.20	26.00
Electric screwdriver	10.00	4.00	20.00
Lamanate floor kit	12.00	4.80	24.00

Sanders

Delta	8.00	3.20	16.00
Palm	7.00	2.80	14.00
Floor	30.00	12.00	60.00
Floor edge	20.00	8.00	40.00
Flap	30.00	12.00	60.00
Orbital	40.00	16.00	80.00

WELDING AND GENERATORS

Generating

Generators			
petrol, 2KVA	20.00	8.00	40.00
petrol, 3KVA	22.50	9.00	45.00
silenced, 2KVA	30.00	12.00	60.00
silenced, 15KVA	180.00	60.00	300.00

PUMPING EQUIPMENT

Pumps

Submersible			
1.25in	11.00	4.40	22.00
2.00in	20.00	8.00	40.00
Puddle pump, electric	25.00	10.00	50.00
Water pump, petrol	25.00	10.00	50.00

	24 hours	Additional 24 hours	Week
	£	£	£

PAINTING AND DECORATING

Surface preparation

Wallpaper stripper, electric	10.00	4.00	20.00
Wallpaper perforator	2.50	1.00	5.00
Scraper, electric	10.00	4.00	20.00

Damp proofing

Injection machine	30.00	12.00	60.00
Woodworm spray attachement	2.50	1.00	5.00
Damp meter	9.00	3.60	18.00

Part Four

GENERAL CONSTRUCTION DATA

General construction data

The metric system

Linear

1 centimetre (cm)	=	10 millimetres (mm)
1 decimetre (dm)	=	10 centimetres (cm)
1 metre (m)	=	10 decimetres (dm)
1 kilometre (km)	=	1000 metres (m)

Area

100 sq millimetres	=	1 sq centimetre
100 sq centimetres	=	1 sq decimetre
100 sq decimetres	=	1 sq metre
1000 sq metres	=	1 hectare

Capacity

1 millilitre (ml)	=	1 cubic centimetre (cm3)
1 centilitre (cl)	=	10 millilitres (ml)
1 decilitre (dl)	=	10 centilitres (cl)
1 litre (l)	=	10 decilitres (dl)

Weight

1 centigram (cg)	=	10 milligrams (mg)
1 decigram (dg)	=	10 centigrams (mcg)
1 gram (g)	=	10 decigrams (dg)
1 decagram (dag)	=	10 grams (g)
1 hectogram (hg)	=	10 decagrams (dag)

Conversion equivalents (imperial/metric)

Length

1 inch	=	25.4 mm
1 foot	=	304.8 mm
1 yard	=	914.4 mm
1 yard	=	0.9144 m
1 mile	=	1609.34 m

Area

1 sq inch	=	645.16 sq mm
1 sq ft	=	0.092903 sq m
1 sq yard	=	0.8361 sq m
1 acre	=	4840 sq yards
1 acre	=	2.471 hectares

Liquid

1 lb water	=	0.454 litres
1 pint	=	0.568 litres
1 gallon	=	4.546 litres

Horse-power

1 hp	=	746 watts
1 hp	=	0.746 kW
1 hp	=	33,000 ft.lb/min

Weight

1 lb	=	0.4536 kg
1 cwt	=	50.8 kg
1 ton	=	1016.1 kg

Conversion equivalents (metric/imperial)

Length

1 millimetre	=	0.03937 inches
1 centimetre	=	0.3937 inches
1 metre	=	1.094 yards
1 metre	=	3.282 ft
1 kilometre	=	0.621373 miles

Area

1 sq millimetre	=	0.00155 sq in
1 sq metre	=	10.764 sq ft
1 sq metre	=	1.196 sq yards
1 acre	=	4046.86 sq m
1 hectare	=	0.404686 acres

Liquid

1 litre	=	2.202 lbs
1 litre	=	1.76 pints
1 litre	=	0.22 gallons

Horse-power
1 watt	=	0.00134 hp
1 kw	=	134 hp
1 hp	=	0759 kg m/s

Weight
1 kg	=	2.205 lbs
1 kg	=	0.01968 cwt
1 kg	=	0.000984 ton

Temperature equivalents

In order to convert Fahrenheit to Celsius deduct 32 and multiply by 5/9.
To convert Celsius to Fahrenheit multiply by 9/5 and add 32.

Fahrenheit	Celsius
230	110.0
220	104.4
210	98.9
200	93.3
190	87.8
180	82.2
170	76.7
160	71.1
150	65.6
140	60.0
130	54.4
120	48.9
110	43.3
100	37.8
90	32.2
80	26.7
70	21.1
60	15.6
50	10.0
40	4.4
30	-1.1
20	-6.7
10	-12.2
0	-17.8

Areas and volumes

Figure	Area	Perimeter
Rectangle	Length x breadth	Sum of sides
Triangle	Base x half of perpendicular height	Sum of sides
Quadrilateral	Sum of areas of contained triangles	Sum of sides
Trapezoidal	Sum of areas of contained triangles	Sum of sides
Trapezium	Half of sum of parallel sides x perpendicular height	Sum of sides
Parallelogram	Base x perpendicular height	Sum of sides
Regular polygon	Half sum of sides x half internal diameter	Sum of sides
Circle	pi x radius2	pi x diameter or pi x 2 x radius

Figure	Surface area	Volume
Cylinder	pi x 2 x radius2 x length (curved surface only)	pi x radius2 x length
Sphere	pi x diameter2	Diameter3 x 0.5236

Weights of materials	kg/m3	kg/m2	kg/m
Aggregate, coarse	1,500		
Ashes	800		
Ballast	600		
Blockboard, standard	940-1000		
Blockboard, tempered	940-1060		
Blocks, natural aggregate			
75mm		160.00	
100mm		215.00	
140mm		300.00	

Weights of materials	kg/m3	kg/m2	kg/m
Blocks, lighweight aggregate			
75mm		60.00	
100mm		80.00	
140mm		112.00	
Bricks, Fletton		1820.00	
Bricks, engineering		2250.00	
Bricks, concrete		1850.00	
Brickwork, 112.5mm		220.00	
Brickwork, 215mm		465.00	
Brickwork, 327.5mm		710.00	
Cement	1,440		
Chalk	2,240		
Chipboard, standard grade	650-750		
Chipboard, flooring grade	680-800		
Clay	1,800		
Concrete	2,450		
Copper pipes, table X			
6mm			0.091
8mm			0.125
10mm			0.158
12mm			0.191
15mm			0.280
18mm			0.385
22mm			0.531
28mm			0.681
35mm			1.133
42mm			1.368
54mm			1.769
Copper pipes, table Y			
6mm			0.117
8mm			0.162
10mm			0.206
12mm			0.251
15mm			0.392
18mm			0.476
22mm			0.697
28mm			0.899
35mm			1.409
42mm			1.700
54mm			2.905

Weights of materials	kg/m3	kg/m2	kg/m
Copper pipes, table Z			
6mm			0.077
8mm			0.105
10mm			0.133
12mm			0.161
15mm			0.203
18mm			0.292
22mm			0.359
28mm			0.459
35mm			0.670
42mm			0.922
54mm			1.334
Flint	2,550		
Gravel	1,750		
Hardcore	1,900		
Hoggin	1,750		
Glass, clear sheet			
3mm		7.50	
4mm		10.00	
5mm		12.50	
6mm		15.00	
10mm		25.00	
12mm		30.00	
15mm		37.50	
19mm		47.50	
25mm		63.50	
Glass, float			
3mm		7.50	
4mm		10.00	
5mm		12.50	
6mm		15.00	
Glass, patterned			
3mm		6.00	
4mm		7.50	
5mm		9.50	
6mm		11.50	
10mm		21.50	
Laminboard	500-700		
Lime, ground	750		

Weights of materials	kg/m3	kg/m2	kg/m
Mild steel flat bars			
25 x 9.53mm			1.910
38 x 9.53mm			2.840
50 x 12.70			5.060
50 x 19.00			7.590
Mild steel round bars			
6mm			0.222
8mm			0.395
10mm			0.616
12mm			0.888
16mm			1.579
20mm			2.466
25mm			3.854
32mm			6.313
40mm			9.864
50mm			15.413
Mild steel square bars			
6mm			0.283
8mm			0.503
10mm			0.784
12mm			1.131
16mm			2.010
20mm			3.139
25mm			4.905
32mm			8.035
40mm			12.554
50mm			19.617
Plaster			
Carlite browning, 11mm thick		7.80	
Carlite tough coat, 11mm thick		7.80	
Carlite bonding, 8mm thick		7.10	
Carlite bonding, 11mm thick		9.80	
Thistle hardwall, 11mm thick		8.80	
Thistle dri-coat, 11mm thick		8.30	
Thistle renovating, 11mm thick		8.80	
Sand	1,600		
Screed, 12.5mm thick		29.00	
Stone, Bath	2,200		
Stone, crushed	1,350		
Stone, Darley Dale	2,400		

Weights of materials	kg/m3	kg/m2	kg/m
Stone, natural	2,400		
Stone, Portland	2,200		
Stone, reconstructed	2,250		
Stone, York	2,400		
Terrazzo, 25mm thick		45.50	
Timber			
Ash	800		
Baltic Spruce	480		
Beech	815		
Birch	720		
Box	960		
Cedar	480		
Ebony	1,215		
Elm	625		
Greenheart	960		
Jarrah	815		
Maple	750		
Pine, Pitchpine	800		
Pine, Red Deal	575		
Pine, Yellow Deal	530		
Sycamore	530		
Teak, African	960		
Teak, Indian	655		
Walnut	495		
Top soil	1,000		
Water	950		
Woodblock flooring			
softwood		12.70	
hardwood		17.60	
Zinc sheeting		4.6	

EXCAVATION AND FILLING

Shrinkage of deposited material

Clay	-10%
Gravel	-7.50%
Sandy soil	-12.50%

Bulking excavated material

Clay	40%
Gravel	25%
Sand	20%

Typical fuel comsumption for plant	**Engine size kW**	**Litres per hour**
Compressors up to	20	4.00
	30	6.50
	40	8.20
	50	9.00
	75	16.00
	100	20.00
	125	25.00
	150	30.00
Concrete mixers up to		
	5	1.00
	10	2.40
	15	3.80
	20	5.00
Dumpers	5	1.30
	7	2.00
	10	3.00
	15	4.00
	20	4.90
	30	7.00
	50	12.00
Excavators	10	2.50
	20	4.50
	40	9.00
	60	13.00
	80	17.00
Pumps	5	1.10
	10	2.10
	15	3.20
	20	4.20
	25	5.50

CONCRETE WORK

Concrete mixes

Mix	Cement t	Sand m3	Aggregate m3	Water litres
1:1:2	0.50	0.45	0.70	208.00
1:1:5:3	0.37	0.50	0.80	185.00
1:2:4	0.30	0.54	0.85	175.00
1:2:5:5	0.25	0.55	0.85	166.00
1:3:6	0.22	0.55	0.85	160.00

BRICKWORK AND BLOCKWORK

Bricks per m2 (brick size 215 x 103.5 x 65mm)

Half brick wall
  stretcher bond   59
  English bond   89
  English garden wall bond   74
  Flemish bond   79

One brick wall
  English bond   118
  Flemish bond   118

One and a half brick wall
  English bond   178
  Flemish bond   178

Two brick wall
  English bond   238
  Flemish bond   238

Metric modular bricks

  200 x 100 x 75mm
    90mm thick   133
    190mm thick   200

200 x 100 x 100mm

90mm thick	50
190mm thick	100
290mm thick	150

300 x 100 x 75mm

90mm thick	44

300 x 100 x 100mm

90mm thick	50

Blocks per m2 (block size 414 x 215mm)

60mm thick	9.9
75mm thick	9.9
100mm thick	9.9
140mm thick	9.9
190mm thick	9.9
215mm thick	9.9

Mortar per m2	**Wirecut m3**	**1 Frog m3**	**2 Frogs m3**
Brick size 215 x 103.5 x 65mm			
Half brick wall	0.017	0.024	0.031
One brick wall	0.045	0.059	0.073
One and a half brick wall	0.072	0.093	0.114
Two brick wall	0.101	0.128	0.155

Brick size 200 x 100 x 75mm	**Solid m3**	**Perforated m3**
90mm thick	0.016	0.019
190mm thick	0.042	0.048
290mm thick	0.068	0.078

Brick size 200 x 100 x 100mm

	Solid m3	Perforated m3
90mm thick	0.013	0.016
190mm thick	0.036	0.041
290mm thick	0.059	0.067

	Solid m3	Perforated m3
Brick size 200 x 100 x 100mm		
90mm thick	0.015	0.018
Block size 440 x 215mm		
60mm thick	0.004	
75mm thick	0.005	
100mm thick	0.006	
140mm thick	0.007	
190mm thick	0.008	
215mm thick	0.009	

MASONRY

Mortar per m2 of random rubble walling	m3
300mm thick wall	0.120
450mm thick wall	0.160
550mm thick wall	0.120

CARPENTRY AND JOINERY

Length of boarding required	m/m2
Board width, 75mm	13.33
Board width, 100mm	10.00
Board width, 125mm	8.00
Board width, 150mm	6.67
Board width, 175mm	5.71
Board width, 200mm	5.00

ROOFING

	Lap mm	Gauge mm	Nr/m2	Battens m/m2
Clay/concrete tiles				
267 x 165mm	65	100	60.00	10.00
	65	98	64.00	10.50
	65	90	68.00	11.30

	Lap mm	Gauge mm	Nr/m2	Battens m/m2
387 x 230mm	75	300	16.00	3.20
	100	280	17.40	3.50
420 x 330mm	75	340	10.00	2.90
	100	320	10.74	3.10
Fibre slates				
500 x 250mm	90	205	19.50	4.85
	80	210	19.10	4.76
	70	215	18.60	4.65
600 x 300mm	105	250	13.60	4.04
	100	250	13.40	4.00
	90	255	13.10	3.92
	80	260	12.90	3.85
	70	263	12.70	3.77
400 x 200mm	70	165	30.00	6.06
	75	162	30.90	6.17
	90	155	32.30	6.45
500 x 250mm	70	215	18.60	4.65
	75	212	18.90	4.72
	90	205	19.50	4.88
	100	200	20.00	5.00
	110	195	20.50	5.13
600 x 300mm	100	250	13.4	4.00
	110	245	13.60	4.08
Natural slates				
405 x 205mm	75	165	29.59	8.70
405 x 255mm	75	165	23.75	6.06
405 x 305mm	75	165	19.00	5.00
460 x 230mm	75	195	23.00	6.00
460 x 255mm	75	195	20.37	5.20
460 x 305mm	75	195	17.00	5.00
510 x 255mm	75	220	18.02	4.60
510 x 305mm	75	220	15.00	4.00

	Lap mm	Gauge mm	Nr/m2	Battens m/m2
560 x 280mm	75	240	14.81	4.12
560 x 280mm	75	240	14.00	4.00
610 x 305mm	75	265	12.27	3.74

Reconstructed stone slates

380 x 250mm	75	150	16.00	3.20

PLASTERING AND TILING

Plaster coverage

	m2 per 1000kg
Carlite browning, 11mm thick	135-155
Carlite tough coat, 11mm thick	135-150
Carlite bonding, 11mm thick	100-115
Thistle hardwall, 11mm thick	115-130
Thistle dri-coat, 11mm thick	135-135
Thistle renovating, 11mm thick	115-125

Tile coverage

	Nr
152 x 152mm	43.27
200 x 200mm	25.00

PLUMBING AND HEATING

Roof drainage

	Area m2	Pipe mm	Gutter mm
One end outlet	15	50	75
	38	68	100
	100	110	150
Centre outlet	30	50	75
	75	68	100
	200	110	150

PAINTING AND WALLPAPERING

Average coverage of paints m2 per litre	Timber	Plastered surfaces	Brickwork
Primer	10-12	9-11	5-7
Undercoat	10-12	11-14	6-8
Gloss	11-14	11-14	6-8
Emulsion	10-12	12-15	6-10

Wallpaper coverage per roll	Rolls nr	Wall height m	Room perimeter m
	4	2.50	8
	5	2.50	9
	5	2.50	10
	6	2.50	11
	6	2.50	12
	7	2.50	13
	7	2.50	14
	8	2.50	15
	8	2.50	16
	8	2.50	17
	9	2.50	18
	10	2.50	19
	10	2.50	20
	10	2.50	21
	11	2.50	22
	11	2.50	23
	12	2.50	24
	13	2.50	25
	13	2.50	26
	14	2.50	27
	5	2.80	8
	5	2.80	9
	5	2.80	10
	7	2.80	11
	7	2.80	12
	7	2.80	13
	8	2.80	14
	8	2.80	15

Rolls nr	Wall height m	Room perimeter m
9	2.8	16
10	2.8	17
10	2.8	18
11	2.8	19
11	2.8	20
12	2.8	21
13	2.8	22
13	2.8	23
14	2.8	24
14	2.8	25
15	2.8	26
15	2.8	27

EXTERNAL WORKS

Blocks/slabs per m2

	nr/m2
200 x 100mm	50.00
450 x 450mm	4.93
600 x 450mm	3.70
600 x 600mm	2.79
600 x 750mm	2.22
600 x 900mm	1.85

Drainage trench widths

	Under 1.5m deep mm	Over 1.5m deep mm
Pipe diameter 100mm	450	600
Pipe diameter 150mm	500	650
Pipe diameter 225mm	600	750
Pipe diameter 300mm	650	800

Volumes of filling for pipe beds (m3 per m)	50mm thick	100mm thick	150mm thick
Pipe diameter 100mm	0.023	0.045	0.068
Pipe diameter 150mm	0.026	0.053	0.079
Pipe diameter 225mm	0.030	0.060	0.090
Pipe diameter 300mm	0.038	0.075	0.113

Volumes of filling for pipe bed and haunching (m3 per m)

Pipe diameter 100mm	0.117
Pipe diameter 150mm	0.152
Pipe diameter 225mm	0.195
Pipe diameter 300mm	0.279

Volumes of filling for pipe bed and surround (m3 per m)

Pipe diameter 100mm	0.185
Pipe diameter 150mm	0.231
Pipe diameter 225mm	0.285
Pipe diameter 300mm	0.391

Part Five

BUSINESS MATTERS

Starting a business

Running a business

Taxation

Starting a business

Most small businesses come into being for one of two reasons – ambition or desperation! A person with genuine ambition for commercial success will never be completely satisfied until he has become self-employed and started his own business. But many successful businesses have been started because the proprietor was forced into this course of action because of redundancy.

Before giving up his job, the would-be businessman should consider carefully whether he has the required skills and the temperament to survive in the highly competitive self-employed market. Before commencing in business it is essential to assess the commercial viability of his intended business because it is pointless to finance a business that is not going to be commercially viable.

In the early stages it is important to make decisions such as: What exactly is the product being sold? What is the market view of that product? What steps are required before the developed product is first sold and where are those sales coming from?

As much information as possible should be obtained on how to run a business before taking the plunge. Sales targets should be set and it should be clearly established how those important first sales are obtained. Above all, do not underestimate the amount of time required to establish and finance a new business venture.

Whatever the size of the business it is important that you put in writing exactly what you are trying to do. This means preparing a business plan that will not only assist in establishing your business aims but is essential if you need to raise finance. The contents of a typical business plan are set out later. It is important to realise that you are not on your own and there are many contacts and advising agencies that can be of assistance.

Potential customers and trade contacts

Many persons intending to start a business in the construction industry will have already had experience as employees. Use all contacts to check the market, establish the sort of work that is available and the current charge-out rates. In the domestic market, check on the competition for prices and services provided. Study advertisements for your kind of work and try to get firm promises of work before the start-up date.

Testing the market

Talk to as many traders as possible operating in the same field. Identify if the market is in the industrial, commercial, local government or in the domestic field. Talk to prospective customers and clients and consider how you can improve on what is being offered in terms of price, quality, speed, convenience, reliability and back-up service.

Business Links

There is no shortage of information about the many aspects of starting and running your own business. Finance, marketing, legal requirements, developing your business idea and taxation matters are all the subject of a mountain of books, pamphlets, guides and courses so it should not be necessary to pay out a lot of money for this information. Indeed, the likelihood is that the aspiring businessman will be overwhelmed with information and will need professional guidance to reduce the risk of wasting time on studying unnecessary subjects.

Business Links are now well established and provide a good place to start for both information and advice. These organisations provide a 'one-stop-shop' for advice and assistance to owner-managed businesses. They will often replace the need to contact Training and Enterprise Councils (TECs) and many of the other official organisations listed below.

Point of contact: telephone directory for address.

Training and Enterprise Councils (TECs)

TECs are comprised of a board of directors drawn from the top men in local industry, commerce, education, trade unions etc, who, together with their staff and experienced business counsellors, assist both new and established concerns in all aspects of running a business. This takes the form of across-the-table advice and also hands-on assistance in management, marketing and finance if required. There are also training courses and seminars available in most areas together with the possibility of grants in some areas.

Point of contact: local Jobcentre or Citizens' Advice Bureau.

Banks

Approach banks for information about the business accounts and financial services that are available. Your local Business Link can advise on how best to find a suitable bank manager and to inform you on what he will require.

Shop around several banks and branches if you are not satisfied at first because managers vary widely in their views on what is a viable business proposition. Remember, most banks have useful free information packs to help business start-up.
Point of contact: local bank manager.

HM Inspector of Taxes

Make a preliminary visit to the local tax office enquiry counter for their publications on income tax and national insurance contributions.

SA/Bk 3	*Self assessment. A guide to keeping records for the self employed*
IR 15(CIS)	*Construction Industry Tax Deduction Scheme*
CWL	*Starting your own business,*
IR 40(CIS)	*Conditions for Getting a Sub-Contractor's Tax Certificate*
NE1	*PAYE for Employers (if you employ someone)*
NE3	*PAYE for new and small Employers*
IR 56/N139	*Employed or Self-Employed. A guide for tax and National Insurance*
CA02	*National Insurance contributions for self employed people with small earnings.*

Remember the onus is on the taxpayer to notify the Inland Revenue that he is in Business and failure to do so may result in the imposition of interest and penalties. Either send a letter or use the form provided at the back of the *'Starting your own business booklet'* to the Inland Revenue National Insurances Contributions office and they will inform your local tax office of the change in your employment status.
Point of contact: telephone directory for address

Inland Revenue National Insurance Contributions Office

Self Employment Services
Customer Accounts Section
Longbenton
Newcastle NE 98 1ZZ

Telephone the Call Centre on 0845 9154655 and ask for the following publications:

CWL2	*Class 2 and Class 4 Contributions for the Self Employed*
CA02	*People with Small Earnings from Self-Employment*
CA04	*Direct Debit - The Easy Way to Pay. Class 2 and Class 3*
CA07	*Unpaid and Late Paid Contributions*
	and for Employers
CWG1	*Employer's Quick Guide to PAYE and NIC Contributions*
CA30	*Employer's Manual to Statutory Sick Pay*

VAT

The VAT office also offer a number of useful publications including;

700	*The VAT Guide*
700/1	*Should I be Registered for VAT?*
731	*Cash Accounting*
732	*Annual Accounting*
742	*Land and Property*

Information about the Cash Accounting Scheme and the introduction of annual VAT returns are dealt with later.

Point of contact: telephone directory for address.

Local authorities

Authorities vary in provisions made for small businesses but all have been asked to simplify and cut delays in planning applications. In Assisted Areas, rent-free periods and reductions in rates may be available on certain industrial and commercial properties. As a preliminary to either purchasing or renting business premises, the following booklets will be helpful:

Step by Step Guide to Planning Permission for Small Businesses and *Business Leases and Security of Tenure*

Both are issued by the Department of Employment and are available at council offices, Citizens' Advice Bureaux and TEC offices. Some authorities run training schemes in conjunction with local industry and educational establishments.

Point of contact: usually the Planning Department - ask for the Industrial Development or Economic Development Officer.

Department of Trade and Industry

The services formally provided by the Department are now increasingly being provided by Business Link. The Department can still, however, provide useful information on available grants for start-ups.

Point of contact: telephone 0207-215 5000 and ask for the address and telephone number of the nearest DTI office and copies of their explanatory booklets.

Department of the Environment

Regulations are now in force are in force relating to all forms of waste other than normal household rubbish. Any business that produces, stores, treats, processes, transports, recycles or disposes of such waste has a 'duty of care' to ensure it is properly discarded and dealt with.

Practical guidance on how to comply with the law (it is a criminal offence punishable by a fine not to) is contained in a booklet *Waste Management: The Duty of Care: A Code of Practice,* obtainable from HMSO Publication Centre, PO Box 276, London SW8 5DT. Telephone 0207-873 9090.

Accountant

The services of an accountant are to be strongly recommended from the beginning because the legal and taxation requirements start immediately and must be properly complied with if trouble is to be avoided later. A qualified accountant must be used if a limited company is being formed but an accountant will give advice on a whole range of business issues including book-keeping, tax planning and compliance to finance raising and will help in preparing annual accounts.

It is worth spending some time finding an accountant who has other clients in the same line of business and is able to give sound advice particularly on taxation and business finance and is not so overworked that damaging delays in producing accounts are likely to arise. Ask other traders whether they can recommend their own accountant. Visit more than one firm of accountants, ask about the fees they charge and how much the production of annual accounts and agreement with the Inland Revenue are likely to cost. A good accountant is worth every penny of his fees and will save you both money and worry.

Solicitor

Many businesses operate without the services of a solicitor but there are a number of occasions when legal advice should be sought. In particular, no-one should sign a lease of premises without taking legal advice because a business can encounter financial difficulty through unnoticed liabilities in its lease. Either an accountant or solicitor will help with drawing up a partnership agreement which all partnerships should have. A solicitor will also help to explain complex contractual terms and prepare draft contracts if the type of business being entered into requires them.

Insurance broker

Policies are available to cover many aspects of business including:

    employer's liability - compulsory if the business has employees
    public liability - essential in the construction industry
    motor vehicles
    theft of stock, plant and money
    fire and storm damage
    personal accident and loss of profits
    keyman cover.

Brokers are independent advisers who will obtain competitive quotations on your behalf. See more than one broker before making a decision - their advice is normally given free and without obligation.
Point of contact: telephone directory or write for a list of local members to:

    The British Insurance Brokers Association
    Consumer Relations Department
    BIBA House
    14 Bevis Marks
    London
    EC3A 7NT (telephone: 0207-623 9043)
                or contact
    The Association of British Insurers
    51 Gresham Street
    London
    EC2V 7HQ (telephone: 0207-600 3333)

who will supply free a package of very useful advice files specially designed for the small business.

The Health and Safety Executive

The Executive operates the legislation covering everyone engaged in work activities and has issued a very useful set of '*Construction Health Hazard Information Sheets*' covering such topics as handling cement, lead and solvents, safety in the use of ladders, scaffolding, hoists, cranes, flammable liquids, asbestos, roofs and compressed gases etc. A pack of these may be obtained free from your local HSE office or The Health & Safety Executive Central Office, Sheffield (telephone: (01142-892345) or HSE Publications (telephone: 01787-881165).

Business plan

As stated before, once the relevant information has been obtained it should be consolidated into a formal business plan. The complexity of the plan will depend in the main on the size and nature of the business concerned. Consideration should be given to the following points.

Objectives

It is important to establish what you are trying to achieve both for you and the business. A provider of finance may be particularly influenced by your ability to achieve short- and medium-term goals set and have confidence in continuing to provide finance for the business. From an individual point of view it is important to establish goals because there is little point in having a business that only serves to achieve the expectations of others whilst not rewarding the would-be businessman.

History

If you already own an existing business then commentary on its existing background structure and history to date can be of assistance. There is no substitute for experience and any existing contacts you have in the construction industry will be of assistance to you. The following points should also be considered for inclusion:

- a brief history of the business identifying useful contacts made

- the development of the business, highlighting significant successes and their relevance to the future
- principal reasons for taking the decision to pursue this new venture
- details of present financing of the business.

Products or services

It is important to establish precisely what it is you are going to sell. Does the product or service have any unique qualities which gives it advantages over your competitors? For example, do you have an ability to react more quickly than they do and are you perceived to deliver a higher quality product or service? A typical business plan would include:

- description of the main products and services
- statement of disadvantages advising how they will be overcome
- details of new products and estimated dates of introduction
- profitability of each product
- details of research and development projects
- after sales support.

Markets and marketing strategy

This section of the business plan should show that thought has been given to the potential of the product. In this regard it can often be useful to identify major competitors and make an overall assessment of their strengths and weaknesses including the following:

- an overall assessment of the market, setting out its size and growth potential
- a statement showing your position within the market
- an identification of main customers and how they compare
- details of typical orders and buying habits
- pricing strategy
- anticipated effect on demand of pricing
- expectation of price movement
- details of promotions and advertising campaigns.

It is important to identify your customers and why they might buy from you. Those entering the domestic side of the business will need to think about the best way to reach potential customers. Are local word-of-mouth recommendations enough to provide reasonable work continuity. If not, what is the most effective method of advertising to reach your customer base?

Remember, advertising is costly. It is a waste of funds to place an advertisement in a paper circulating in areas A, B, C & D if the business only covers area A.

Research and development

If you are developing a product or a particular service, then an assessment should be made on what stage it is at and what further finance is required to complete it. It may also be useful to make an assessment on the vulnerability of the product or service to innovations being initiated by others.

Basis of operation

Detail what facilities you will require in order to carry on your trade in the form of property, working and storage areas, office space, etc. An assessment should also he made on the assistance you will require from others. Your business plan might include:

-   a layman's guide to the process or work
-   details of facilities, buildings and plant
-   key factors affecting production such as yields and wastage
-   raw material demand and usage.

Management

This section is one of the most important because it demonstrates the capability of the would-be businessman. The skills you need will cover production, marketing, finance and administration. In the early stages you may be able to do this yourself but as the business grows it may be required to develop a team to handle these matters. The following points should be considered for inclusion in the plan:

-   set out age, experience and achievements
-   state additional management requirements in the future and how they are to be met
-   identify current weaknesses and how they will be overcome
-   state remuneration packages and profit expectations
-   give detailed CVs in appendices.

Advertising and retraining may be required in order to identify and provide suitable personnel where expertise and experience are lacking.

Financial information

It is important to detail, if any, the present financial position of your business and the budgeted profit and loss accounts, cash flows and balance sheets. These integrated forecasts should be prepared for the next twelve months at monthly intervals and annually for the following two years.

If the forecasts are to be reasonably accurate then the businessman must make some early decisions about:

-   premises where the business will be based, the initial repairs and alterations that might he required and an assessment of the total cost
-   which plant, equipment and transport is needed, whether it is to be leased or purchased and what the cost will be?
-   how much stock of materials, if any, should be carried - the bare minimum only should be acquired so reliable suppliers should be found
    what will be the weekly bills for overheads, wages and the proprietor's living costs?
-   what type of work is going to be undertaken, how much profit can realistically be obtained
-   how often are invoices to be presented?

Your business plan should include the following information:

-   explanation of how sales forecasts are prepared
-   levels of production
-   details of major variable overheads and estimates
-   assumptions in cash flow forecasting, inflation and taxation.

Finance required and its application

The financial details given above should produce an accurate assessment of the funds required to finance the business. It is important to distinguish between those items that require permanent finance and those that will eventually be converted to cash because it is not usually advisable to finance long-term assets with personal equity.

Working capital such as stock and debtors can usually be obtained by an overdraft arrangement but your accountant or bank will advise you on this.

Executive summary

Although it is prepared last, this summary will be the first part of your business plan. Remember that business plans are prepared for busy people and their decision on finance may be based solely on this section. It should cover two or three pages and deal with the most important aspects and opportunities in your plan. Here are some of the main headings:

- key strategies
- finance required and how it is to be used
- management experience
- anticipated returns and profits
- markets.

The appendices should include:

- CVs of key personnel
- organisation charts
- market studies
- product advertising literature
- professional references
- financial forecasts
- glossary of terms.

If you feel any additional information should be provided in support of your proposal, then this is usually best included in the appendices.

Follow up

Please remember that once your plan is prepared, it is important to re-examine it regularly and update the forecasts and financial information. This is a working document and can be an important tool in running the business.

Sources of finance

Personal funds

Finance, like charity, often begins at home and a would-be businessman should make a realistic assessment of his net worth, including the value of his house after deducting the mortgage(s) outstanding on it, savings, any car or van owned and any sums which his family are prepared to contribute but deducting any private borrowings which will come due for payment. The whole of these funds may not be available (for instance, money which has been loaned to a friend or relative who is known to be unable to repay at the present time).

It may not be desirable that all capital should all be put at risk on a business venture so the following should be established:

- how much cash you propose to invest in the business
- whether the family home will be made available for any business borrowing
- state total finance required
- how finance is anticipated being raised
- interest and security to be provided
- expected return on investment.

Whilst it may be wise not to pledge too much of the family assets, it has to be remembered that the bank will be looking closely at the degree to which the proprietor has committed himself to the venture and will not be impressed by an application for a loan where the applicant is prepared to risk only a small fraction of his own resources.

Having decided how much of his own funds to contribute, the businessman can now see the level of shortfall and consider how best to fill it. Consideration should be given to partners where the shortfall is large and particularly when there is a need for heavy investment in fixed assets such as premises and capital equipment. It may be worthwhile starting a limited company with others also subscribing capital and to allow the banks to take security against the book debts.

Banks

The first outside source of money to which most businessmen turn is the bank and here are a few guide lines on approaching a bank manager:

- present your business plan to him; remember to use conservative estimates which tend to understate rather than overstate the forecast sales and profits
- know the figures in detail and do not leave it to your accountant to explain them for you. The bank manager is interested in the businessman not his advisers and will be impressed if the businessman demonstrates a grasp of the financing of his business
- understand the difference between short and long term borrowing
- ask about the Government Loan Guarantee Scheme if there is a shortage of security for loans. The bank may be able to assist, or depending on certain conditions being met, the government may guarantee a certain percentage of the bank loan.

Remember the bank will want their money back, so bank borrowings are usually required to be secured by charges on business assets. In start-up situations, personal guarantees from the proprietors are normally required. Ensure that if these are given they are regularly reviewed to see if they are still required.

Enterprise Investment Scheme - business angels

If an outside investor is sought in a business he will probably wish to invest within the terms of the Enterprise Investment Scheme which enables him to gain income tax relief at 20% on the amount of his investment. Additionally, any investment can be used to defer capital gains tax. The rules are complex and professional advice should always be sought.

Hire purchase/leasing

It is not always necessary to purchase assets outright required for the business and leasing and hire purchase can often form an integral part of a business's medium-term finance strategy.

Venture capital

In addition, there are a number of other financial institutions in the venture capital market that can help well established businesses, usually limited companies, who wish to expand. They may also assist well-conceived start-ups. They will provide a flexible package of equity and loan capital but only for large amounts, usually sums in excess of £150,000 and often £250,000.

Usually the deal involves the financial institution having a minority interest in the voting share capital and a seat on the board of the company. Arrangements for the eventual purchase of the shares held by the finance company by the private shareholders are also normally incorporated in the scheme.

The Royal Jubilee and Princes Trust

These trusts through the Youth Business Initiative provide bursaries of not more than £1,000 per individual to selected applicants who are unemployed and age 25 or over. Grants may be used for tools and equipment, transport, fees, insurance, instruction and training but not for working capital, rent and rates, new materials or stock. They operate through a local representative whose name and address may be ascertained by contacting the Prince's Youth Business.
Point of contact: telephone 0207-321 6500.

The Business Start-up Scheme

This is an allowance of £50 per week, in addition to any income made from your business, paid for twenty weeks. To qualify you must be at least 18 and under 65, work at least 36 hours per week in the business and have been unemployed for at least six months or fall into one of the other categories; disabled, ex-HMS or redundant.

   The first step is to get the booklet on the subject from your local Jobcentre or TEC that includes details on how and where to apply. Once in receipt of the enterprise allowance, you will also have the benefit of advice and assistance from an experienced businessman from your TEC. All the initial counselling services and training courses are free.

Running a business

Many businesses are run without adequate information being available to check trend in their vital areas, e.g. marketing, money and managerial efficiency. It is essential to look critically at all aspects of the business to maximise profits and reduce inefficiency. Regular meaningful information is required on which management can concentrate. This will vary according to the proprietor's business but will often concentrate on debtors, creditors, cash, sales and orders.

Proprietors often have the feeling that the business should be 'doing better' but are unable to identify what is going wrong. Sometimes there is the worrying phenomenon of a steadily increasing work programme coupled with a persistently reducing bank balance or rising overdraft. Here are some useful ways of checking the position and identifying problem areas are given below.

Marketing

Throughout his business life the entrepreneur should continuously study the methods and approach of his competitors. A shortcoming frequently found in ailing concerns is that the proprietor thinks he knows what his customers want better than they do.

The term 'market research' sounds both difficult and expensive but a very simple form of it can be done quite effectively by the businessman and his sales staff. Existing and prospective customers should be approached and asked what they want in terms of price, quality, design, payment terms, follow-up service, guarantees and services.

The initial approach might be by a leaflet or letter followed by a personal call. As an on-going part of management, all staff with customer contact should be encouraged to enquire about and record customer preferences, complaints, etc. and feed it back to management.

Other sources of information can be trade and business journals, trade exhibitions, suppliers and representatives from which information about trends, new techniques and products can be obtained and studied. Valuable information can also be gained from studying competitors and the following questions should be asked:

- what do they sell and at what prices

- what inducements do they offer to their customers, e.g. credit facilities, guarantees, free offers and discounts
- how do they reach their customers - local/national advertising, mail shots, salesmen, local radio and TV
- what are the strongest aspects of their appeal to customers and have they any weaknesses.

The businessman should apply all the information gathered from customers and competitors to his own services with a view to making sure he is offering the right product at the right price in the most attractive way and in the most receptive market.

In a small business where the proprietor is also his own salesman he must give careful thought on how he can best present his product and himself. For instance, if he is working solely within the construction industry his main problems are likely to centre on getting a C1S6 Certificate and using trade contacts to get sub-contract work.

However, for those who serve the general public, presentation can be a vital element in getting work. The customer is looking for efficiency, reliability and honesty in a trader and quality, price and style in the product. To bring out these facets in discussion with a potential customer is a skilled task. A short course on marketing techniques could pay handsome dividends. The Business Link will give the names and addresses of such courses locally.

Financial control

Unfortunately, some unsuccessful firms do not seek financial advice until too late when the downward trend cannot be halted. Earlier attention to the problems may have saved some of them so it is important to recognise the tell-tale signs. There are some tests and checks that can be done quite easily.

Cashflow

Cashflow is the lifeblood of the business and more businesses fail through lack of cash than for any other reason. Cash is generated through the conversion of work into debtors and then into payment and also through the deferral of the payment of supplies for as long a period that can be negotiated. The objective must be to keep stock, work in progress and debts to a minimum and creditors to a maximum.

Debtor days

This is calculated by dividing your trade debtors by annual sales and multiplying by 365. This shows the number of days' credit being afforded to your customers and should be compared both with your normal trade terms and the previous month's figures. Normal procedures should involve the preparation of a monthly-aged list of debtors showing the name of the customer, the value and to which month it relates.

The oldest and largest debtors can be seen at a glance for immediate consideration of what further recovery action is needed. The list may also show over-reliance on one or two large customers or the need to stop supplying a particularly bad payer until his arrears have been reduced to an acceptable level. Consideration should be given to making up bills to a date before the end of the month and making sure the accounts are sent out immediately, followed by a statement four weeks later.

Consider giving discounts for prompt payment. If all else fails, and legal action for recovery is being contemplated, call at the County Court and ask for their leaflets.

Stockturn

The level of stock should be kept to a minimum and the number of days' stock can be calculated by dividing the stock by the annual purchases and multiplying by 365. A worsening trend on a month-by-month basis shows the need for action. It is important to make regularly a full inventory of all stock and dispose of old or surplus items for cash. A stock control procedure to avoid stock losses and to keep stock to a minimum should be implemented.

Profitability

Whilst cash is vital in the short-term, profitability is vital in the medium-term. The two key percentage figures are the gross profit percentage and the net profit percentage. Gross profit is calculated by deducting the cost of materials and direct labour from the sales figures whilst net profit is arrived at after deducting all overheads. Possible reasons for changes in the gross profit percentage are:

- not taking full account of increases in materials and wages in the pricing of jobs
- too generous discount terms being offered

-   poor management, over-manning, waste and pilferage of materials
-   too much down-time on equipment which is in need of replacement.

If net profit is deteriorating after the deduction of an appropriate reward for your own efforts, including an amount for your own personal tax liability, you should review each item of overhead expenditure in detail asking the following questions:

-   can savings be made in non-productive staff?
-   is sub-contracting possible and would it be cheaper?
-   have all possible energy-saving methods been fully explored?
-   do the company's vehicles spend too much time in the yard and can they be shared or their number reduced?
-   is the expenditure on advertising producing sales - review in association with 'marketing' above?

Over-trading

Many inexperienced businessmen imagine that profitability equals money in the bank and in some cases, particularly where the receipts are wholly in cash, this may be the case. But often increased business means higher stock inventories, extra wages and overheads, increased capital expenditure on premises and plant all of which require short-term finance.

Additionally, if the debtors show a marked increase as the turnover rises, the proprietor may find to his surprise that each expansion of trade reduces rather than increases his cash resources and he is continually having to rely on extensions to his existing credit.

The business, which had enough funds for start-up, finds it does not have sufficient cash to run at the higher level of operation and the bank manager may be getting anxious about the increasing overdraft. It is essential for those who run a business that operates on credit terms to be aware that profitability does not necessarily mean increased cash availability. Regular monthly management information on marketing and finance as described in this chapter will enable over-trading to be recognised and remedial action to he taken early.

If the situation is appreciated only when the bank and other creditors are pressing for money, radical solutions may be necessary such as bringing in new finance, sale and leaseback of premises, a fundamental change in the terms of trade or even selling out to a buyer with more resources. Professional help from the firm's accountant will be needed in these circumstances.

Break-even point

The costs of a business may be divided into two types - variable and fixed. *Variable costs* are those which increase or decrease as the volume of work goes up or down and include such items as materials used, direct labour and power machine tools. *Fixed costs* are not related to turnover and are sometimes called fixed overheads. They include rent, rates, insurance, heat and light, office salaries and plant depreciation. These costs are still incurred even though few or no sales are being made.

Many small businessmen run their enterprises from home using family labour as back-up; they mainly sell their own labour and buy materials and hire plant only as required. By these means they reduce their fixed costs to a minimum and start making profits almost immediately. However, larger firms that have business premises, perhaps a small workshop, an office and vehicles, need to know how much they have to sell to cover their costs and become profitable.

In the case of a new business it is necessary to estimate this figure but where annual accounts are available a break-even chart based on them can be readily prepared. Suppose the real or estimated figures (expressed in £000s) are:

	%	£
Sales	100	400
Variable costs	66	265
Gross profit	34	135
Fixed costs	13	50
Net Profit	21	85

Break-even point $=$ $\dfrac{50 \text{ divided by (1 less variable costs \%)}}{\text{sales}}$

$=$ 50 divided by (1 less 0.6625)
$=$ 50 divided by 0.3375
$=$ £148 (thousand)

In practice things are never quite as clear cut as the figures show, but nevertheless this is a very useful tool for assessing not only the break-even point but also the approximate amount of loss or profit arising at differing levels of turnover and also for considering pricing policy.

Taxation

INCORPORATION

The first decision usually required to be made from a taxation point of view is which trading entity to adopt. The options available are set out below.

Sole trader

A sole trader is a person who is in business on his own account. There is no statutory requirement to produce accounts nor is there a necessity to have them audited. A sole trader may, however, be required to register for PAYE and VAT purposes and maintain records so that Income Tax and VAT returns can be made. A sole trader is personally liable for all the liabilities of his business.

Partnership

A partnership is a collection of individuals in business on their own account and whose constitution is generally governed by the Partnership Act 1890. It is strongly recommended that a partnership agreement is also established to determine the commercial relationship between the individuals concerned.

The requirements in relation to accounting records and returns are similar to those of sole traders and in general a partner's liability is unlimited.

Limited company

This is the most common business entity. Companies are incorporated under the Companies Act 1985 which requires that an annual audit is carried out for all companies with a turnover in excess of £1,000,000 or a review if the turnover is less than £1,000,000 and that accounts are filed with the Companies Registrar. Generally an individual shareholder's liability is limited to the amount of the share capital he is required to subscribe.

Advantages

In view of the problems and costs of incorporating an existing business, it is important to try and select the correct trading medium at the commencement of operations. It is not true to say that every business should start life as a company. Many businesses are carried on in a safe and efficient manner by sole traders or partnerships. Whilst recognising the possible commercial advantages of a limited company, taxation advantages exist for sole traderships and partnerships such as income tax deferral and National Insurance saving. No decision should be taken without first seeking professional advice.

The benefit of limited liability should not be ignored although this can largely be negated by banks seeking personal guarantees. In addition, it may be easier for the companies to raise finance because the bank can take security on the debts of the company that could be sold in the future, particularly if third-party finance has been obtained in the form of equity.

Self-assessment

From the tax year 1996/97 the burden of assessing tax shifted from the Inland Revenue to the individual tax payer. The main features of this system are as follows:

- the onus is on the taxpayer to provide information and complete returns
- tax will be payable on different dates
- the taxpayer has a choice: he can calculate his tax liability at the same time as making his return and this will need to be done by 31st January following the end of the tax year. Alternatively, he can send in his tax return before 30 September and the Inland Revenue will calculate the tax to be paid on the following 31 January
- the important aspect to the system is that if the return is late, or the tax is paid late, there will be automatic penalties and/or surcharges imposed on the taxpayer.

Tax correspondence

Businessmen do not like letters from the Inland Revenue but they should resist the temptation to tear them up or put them behind the clock and forget about them. All Tax Calculations and Statements of Account should be checked for accuracy immediately and any queries should be put your accountant or sent to the Tax District that issued the document.

Keep copies of all correspondence with the Inland Revenue. Letters can be mislaid or fail to be delivered and it is essential to have both proof of what was sent as well as a permanent record of all correspondence.

Dates tax due

Income Tax
Payments on account (based on one half of last year's liability) are due on 31 January and 31 July. If these are insufficient there is a balancing payment due on the following 31 January – the same day as the tax return needs to be filed. For example:

for the year 2000/01   Tax due £5,000 (1999/00 was £4,000)
                         First payment on account of £2,000 is due on 31.01.01
                         Second payment on account of £2,000 is due on 31.07.01
                         Balancing payment of £1,000 is due on 31.01.02

Note that on 31.01.02, the first payment on account of £2,500 will fall due for the next tax year 2001/02.

Tax in business

Spouses in business
If spouses work in the business, perhaps answering the phone, making appointments, writing business letters, making up bills and keeping the books, they should be properly remunerated for it. Being a payment to a family member, the Inspector of Taxes will be understandably cautious in allowing remuneration in full as a business expense. The payment should be:

- actually paid to them, preferably weekly or monthly and in addition to any housekeeping monies
- recorded in the business book
- reasonable in amount in line with their duties and the time spent on them.

If the wages paid to them exceed £76.00 per week, Class 1 employer's and employee's NIC becomes due and if they exceed £4,385 p.a. (assuming they have no other income) PAYE tax will also be payable.

It should also be noted that once small businesses are well established and the spouses' earnings are approaching the above limits, consideration may be given to bringing them is as a partner. This has a number of effects:

- there is a reduced need to relate the spouse's income (which is now a share of the profits) to the work they do
- they will pay Class 2 and Class 4 NIC instead of the more costly Class I contributions and PAYE will no longer apply to their earnings
- but remember that, as partners, they have unlimited liability.

Premises

Many small businessmen cannot afford to rent or buy commercial premises and run their enterprises from home using part of it as an office where the books and vouchers, clients' records and trade manuals are kept and estimates and plans are drawn up. In these circumstances, a portion of the outgoings on the property may be claimed as a business expenses. An accountant's advice should be sought to ensure that the capital gains tax exemption that applies on the sale of the main residence is not lost.

Fixed Profit Car Scheme

It may be advantageous to calculate your car expenses using a fixed rate per business mile. A condition is that your annual turnover is below the VAT threshold (currently £52,000). Ask your accountant about this. A proper record of business mileage must be kept.

Vehicles

Car expenses for sole traders and partners are usually split on a fractional mileage basis between business journeys, which are allowable, and private ones, which are not, and a record of each should he kept. If the business does work only on one or two sites for only one main contractor, the inspector may argue that the true base of operations is the work site not the residence and seek to disallow the cost of travel between home and work. It is tax-wise and sound business practice to have as many customers as possible and not work for just one client.

Business entertainment

No tax relief is due for expenditure on business entertainment and neither is the VAT recoverable on gifts to customers, whether they are from this country or overseas. However, the cost of small trade gifts not exceeding £10 per person per annum in value is still admissible provided that the gift advertises the business and does not consist of food, drink or tobacco.

Income tax

Personal allowances
The current personal allowance for a single person is £4,385. The personal allowance for people aged 65 to 74 and over 75 years are £5,790 and £6,050 respectively. The married couple's allowance was withdrawn on 5 April 2000, except for those over 65 on that date.

Taxation of husband and wife
A married woman is treated in much the same way as a single person with her own personal allowance and basic rate band. Husband and wife each make a separate return of their own income and the Inland Revenue deals with each one in complete privacy; letters about the husband's affairs will be addressed only to him and about the wife's only to her unless the parties indicate differently.

Rates of tax

Tax is deducted at source from most banks and building societies accounts at the rate of 20%. The rates of tax for 2000/01 are as follows:

Lower rate: 10% on taxable income up to £1,500
Basic rate: 22% on taxable income between £1,521 and £28,400
Higher rate: 40% on taxable income over £28,400

Dividends carry a 10% non-repayable tax credit. Higher rate taxpayers pay a further tax on dividends of 22.5%.

Mortgage interest relief

This is no longer available after 5 April 2000.

Business losses

These are allowed only against the income of the person who incurs the loss. For example, a loss in the husband's business cannot be set against the wife's income from employment.

Joint income

In the case of joint ownership by husband and wife of assets which yield income, such as bank and building society accounts, shares and rented property, the Inland Revenue will treat the income as arising equally to both and each will pay tax on one half of the income. If, however, the asset is owned in unequal shares or one spouse only and the taxpayer can prove this, then the shares of income to be taxed can be adjusted accordingly if a joint declaration is made to the tax office setting out the facts.

Capital Gains Tax

Where an asset is disposed of, the first £7,200 of the gain is exempt from tax. In the case of husbands and wives, each has a £7,200 exemption so if the ownership of the assets is divided between them, it is possible to claim exemption on gains up to £14,400 jointly in the tax year. Any remaining gain is chargeable as though it were the top slice of the individual's income; therefore according to his or her circumstances it might be charged at 10%, 22% or 40%.

Retirement relief may be due on the disposal of business assets after the age 50, but is gradually being withdrawn. The maximum relief against capital gains for 2000/01 is £150,000 plus one half the gains between £150,000 and £600,000. A businessman selling a business asset or contemplating retiring or selling the business when aged 50 or over, should consult his accountant *before* taking any steps and *before* changing his working pattern (e.g. going part time).

Self-employed NIC rates (from 6 April 2000)

Class 2 rate
Charged at £2.00 per week. If earnings are below £3,825 per annum averaged over the year, ask the DSS about 'small income exception'. Details are in leaflet CA02.

Class 4 rate

Business profits up to £4,385 per annum are charged at NIL. Annual profits between £4,385 and £27,820 are charged at 7% of the profit. There is no charge on profits over £27,820 so the maximum amount of Class 4 contributions is £1,640.45. Class 4 contributions are collected by the Inland Revenue along with the income tax due. For the year ending 31 March 2001, Corporation Tax is charged at these rates:

£1 to £10,000	= 10%
£10,001 to £50,000	= 22.5%
£50,001 to £300,000	= 20%
£300,001 to £1,500,000	= 32.5%
over £1,500,000	= 30%

Companies can carry back trading losses for up to 3 years.

Capital allowances (depreciation) rates

Plant and machinery:	25% (40% first year allowance is available for certain small businesses).
Business motor cars - cost up to £12,000:	25%
- cost over £12,000:	£3,000 (maximum)
Industrial buildings:	4%
Commercial and industrial buildings in Enterprise Zones:	100%
Computers and software equipment	100%

THE CONSTRUCTION INDUSTRY TAX DEDUCTION SCHEME

General

The new Construction Industry Tax Deduction Scheme is known as the 'CIS' scheme and replaced the old '714' scheme. As the scheme operates whenever a contractor makes a payment to a sub-contractor, the businessman should visit his local income tax enquiry office and obtain copies of the Revenue booklet IR 14/15 (CIS) and leaflet IR 40 which will explain the conditions under which the Revenue will issue a registration card or (CIS6) certificate and precisely when the scheme applies.

Everyone who carries out work in the Construction Industry Scheme must hold a registration card (CIS4), or a tax certificate (CIS6). Certain larger companies use a special certificate (CIS5).

If the sub-contractor has a registration card but does not hold a valid tax certificate (CIS6) issued to him by the Inland Revenue, then the contractor *must* deduct 18% tax from the whole of any payment made to him (excluding the cost of any materials) and to account to the Inland Revenue for all amounts so withheld.

To enable the subcontractor to prove to the Inspector of Taxes that he has suffered this tax deduction the contractor must complete 3 part tax payment voucher (CIS25) showing the amount withheld. These vouchers must be carefully filed for production to the Inspector after the end the tax year along with the tax return. Any tax deducted in this way over and above the sub-contractor's agreed liability for the year will be repaid by the Inland Revenue. If, however, he holds a (CIS6) certificate the payment may be made in full without deducting tax.

A small business that does work only for the general public and small commercial concerns, is outside the scheme and does not need a certificate to trade. If, however, it engages other contractors to do jobs for it, the business would have to register under the scheme as a contractor and deduct tax from any payment made to a sub-contractor who did not produce a valid (CIS6) certificate. If in doubt, consult your accountant or the Inland Revenue direct.

VAT

The general rule about liability to register for VAT is given in the VAT office notes. It is possible to give here only a brief outline of how the tax works. The rules that apply to the construction industry are extremely complex and all traders must study *The VAT Guide* and other publications.

Registration for VAT is required if, at the end of any month the value of taxable supplies in the last 12 months exceeds the annual threshold or if there are reasonable grounds for believing the value of the taxable supplies in the next 30 days will exceed the annual threshold.

Taxable supplies include any zero-rated items. The annual threshold is £53,000. The amount of tax to be paid is the difference between the VAT charged out to customers *(output tax)* and that suffered on payments made to suppliers for goods and services *(input tax)* incurred in making taxable supplies. Unlike income tax there is no distinction in VAT for capital items so that the tax charged on the purchase of, for example, machinery, trucks and office furniture, will normally be reclaimable as Input Tax.

VAT is payable in respect of three monthly periods known as 'tax periods'. You can apply to have the group of tax periods that fits in best with your financial year. The tax must be paid within one month of the end of each tax period. Traders who receive regular repayments of VAT can apply to have them monthly rather than quarterly. Not all types of goods and services are taxed at 17.5% (i.e. the standard rate). Some are exempts and others are zero-rated.

Zero-rated

This means that no VAT is chargeable on the goods or services, but a registered trader can reclaim any *input* tax suffered on his purchases. For instance a builder pays VAT on the materials he buys to provide supplies of constructing but if he is constructing a new dwelling house, this is zero rated. He may reclaim this VAT or set it off against any VAT due on standard rated work.

Exempt

Supplies that are exempt are less favourably treated than those that are zero-rated. Again no VAT is chargeable on the goods or services but the trader cannot reclaim any *input* tax suffered on his purchases.

Standard-rated

All work which is not specifically stated to be zero-rated or exempt is standard-rated, i.e. VAT is chargeable at the current rate of 17.5% and the trader may deduct any *input* tax suffered when he is making his return to the Customs and Excise. If for any reason a trader makes a supply and fails to charge VAT when he should have done so (e.g. mistakenly assuming the supply to be zero rated), he will have to account for the VAT himself out of the proceeds. If there is any doubt about the VAT position, it is safer to assume the supply is standard rated, charge the appropriate amount of VAT on the invoice and argue about it later.

Time of supply

The *time* at which a supply of goods or services is treated as taking place is important and is called the 'tax point'. VAT must be accounted for to the Customs and Excise at the end of the accounting period in which this 'tax point' occurs. For the supply of goods which are 'built on site', the 'basic tax point' is the date the goods are made available for the Customer's use, whilst for *services* it is normally the date when all work except invoicing is completed.

However, if you issue a tax invoice or receive a payment before this 'basic tax point' then that date becomes a tax point. In the case of contracts providing for stage and retention payments the tax point is either the date the tax invoice is issued or when payment is received, whichever is the earlier.

All the requirements apply to sub-contractors and main contractors and it should be noted that, when a contractor deducts income tax from a payment to a sub-contractor (because he has no valid CIS6, VAT is payable on the full gross amount *before* taking off the income tax.

Annual accounting

It is possible to account for VAT other than on a specified three month period. Annual accounting provides for nine equal instalments to be paid by direct debit with annual return provided with the tenth payment. Note that the Customs and Excise have announced changes to the VAT Annual Accounting that will be introduced on 1 April 1996. The new scheme will continue to be voluntary and will particularly benefit businesses with turnover of below £100,000, although the scheme remains available to businesses with an annual turnover of less than £300,000.

Cash accounting

If turnover is below a specified limit, currently £350,000 (£600,000 from 6 April 2001), a taxpayer may account for VAT on the basis of cash paid and received. The main advantages are automatic bad debt relief and a deferral of VAT payment where extended credit is given.

Bad debts

Relief is available for debts over 6 months.

Index

Access and site equipment,
210-212
Accountant, 243
Alterations and repairs, xxi,
164-187
Asphalt generally
damp-proofing, 75-76
flooring, 76-77
roofing, 77-78
tanking, 75-76
Asphalt work, 75-78

Balustrades, 60
Banks, 240-241
Bearers, 45-46
Beds, concrete, 161
Bitumen macadam paving, 162
Blockwork, xix, 30-34
Boilers
gas, 110
oil, 111
Breaking and demolition tools,
213-215
Break-even point, 257
Brick paving, 163
Brickwork, xix, 20-30
Brickwork and blockwork, xix,
20-36,173-176, 194-195,
228-230
Business links, 240
Business matters, 239-268
Business plan, 245-249

Capital Gains Tax, 264
Carpentry and joinery, xix-xx,
41-59, 180-182, 195-199

Cashflow, 254
Cast iron
gutters, 95-96
rainwater pipes, 94-95
soil pipes, 98-99
Cavity wall insulation, 36
Ceramic tiling, 88-89
Chainlink fencing, 157-158
Chestnut fencing, 158
Cistern, 106-107
Clear float glass, 117-119
Clear laminated glass, 123-124
Close boarded fencing, 158
Cold water storage tanks, 106-107
Concrete and cutting equipment,
207-209
Concrete beds, 161
Concrete finishes, 18
Concrete mixes, 228
Concrete work, xix, 11-19, 192
Construction Industry
Deduction Scheme, 265-266
Copings, 19
Copper pipework
capillary joints, 100-103
compression joints, 104-106
Copper sheet coverings, 72-73
Cylinders, 107-109

Damp proof courses, 34-35
Damp proofing, 75-76
Debtor days, 255
Demolition, xix, 3-4
Demolition, excavation and filling,
3-10

Department of the
   Environment, 243
Department of Trade
   and Industry, 243
Disposal, 7-8
Doors, 52-53
Door frames and linings, 53-55
Drainage, 148-157, 202-204,

Earthwork support, 7
Excavation and filling, xix, 5-6
   191, 226-227
Excavation and site plant, 209
Expansion joints, 161-162
External works, xxi, 148-163,
   234-235

Fascias, 45
Fencing
   chainlink, 157-158
   chestnut, 158
   close boarded, 158
   strained wire, 158
Fibre-cement slating, 65-66
Filling, xix, 9-10
Filling openings, 167
Finance
   banks, 250-251
   Business Start-up Scheme, 252
   Enterprise Investment
      Scheme, 251
   hire purchase/leasing, 251
   personal funds, 250
   Royal Jubilee and Princes
      Trust, 252
   venture capital, 251-252
Finishings, ix, 79-89, 182-183, 198
Fixed Profit Car Scheme, 262
Flooring, 48-49, 76-77

Floor joists, 41, 42-43
Floor tiling, 86-88
Floor, wall and ceiling finishings,
   79-89
Flue pipe, 111-112
Forming openings, 165-166
Formwork, 13-16
Fuel comsumption, 227

Gas boilers, 110
General construction data, 219-235
Georgian glass
   wired, 121
   wired polished, 122-123,
Glass
   clear float, 117-119
   clear laminated safety, 123-124
   Georgian wired, 121
   Georgian wired polished,
      122-123
   white patterned, 119-121
Glazing, xxi, 117-124, 185
Granolithic screed, 81-82
Gravel paving, 162
Gutter boards, 44-45

Health and Safety
   Executive, 245
Hardwood windows, 51
Holes and chases, 92-93
Hot water tanks, 107

Inspector of Taxes, 241
Insulation, 57, 109-110
Insurance broker, 244-245
Ironmongery, 57-59

Joinery, see Carpentry and
   Joinery
Joints, 18

Kerbs and bearers, 42, 43-44
Kerbs and edgings, 159-160
Kitchen fittings, 56

Labour, xv-xvii
Lathing, 85
Lead sheet coverings, 69-71
Limited company, 259
Linings, 49-50
Linoleum, 88
Lintels
  concrete, 18-19
  steel, 60
Local authorities, 242

Manholes, 155-157
Marketing, 253
Masonry, xix, 37-40, 194-195, 230
Materials, xv
Material handling, 212
Metal fixings, 46-47
Metalwork, xx, 60-61

National Insurance
  contributions, 264-265
Natural slating, 66-68

Oil boilers, 111
Oil storage tanks, 112-113
Overflows, 99
Overheads and profit, xvii
Over-trading, 256

Painting,
  generally, xxi, 125-145
  186, 200-201, 233
  internally, 125-136
  externally, 136-145,
Partnership, 259

Pavings
  bitumen macadam, 162
  brick, 163
  gravel, 162
  precast concrete, 163
Plasterboard, 86
Plasterwork, 83-84, 232
Plumbing and heating, xx, 90-116,
  183-185, 198-200, 232
Pointing, 36
Precast concrete
  copings, 19
  kerbs, 160
  lintels, 18
  pavings, 163
Premises, 262
Profitability, xvii, 255-256
Pumping equipment, 215
PVC-U
  rainwater gutters, 95
  rainwater pipes, 94
  soil pipes, 97-98
  windows, 51

Quarry tiling, 86-87

Radiators, 114-116
Radiator valves, 104
Rainwater goods, 94-96
Rainwater gutters
  cast iron, 95-96
  PVC-U, 95
Rainwater pipes
  cast iron, 94-95
  PVC-U, 94
Reconstructed slating, 68-69
Reinforcement
  bars, 17
  fabric, 17

Roof joists, 41, 43
Roof tiling, 62-65
Rubble walling, 37-38

Sanitary fittings, 90-92
Screeds
  cement and sand, 79-81
  granolithic, 81-82
Screens, 173
Self-assessment, 260
Sheet coverings
  copper, 72-73
  lead, 69-71
Shoring, 164
Slating
  fibre-cement, 65-66
  natural, 66-68
  reconstructed, 68-69
Softwood windows, 50-51
Soil pipes
  cast iron, 98-99
  PVC-U, 97-98
Sole trader, 259
Solicitor, 244
Spot items, 173-187
Stairs, 55-56
Steel lintels, 60
Stop valves, 104, 106
Strained wire fencing, 158
Strutting, 42, 44
Surface treatments, 10
Sub-bases, 160

Tanking, 75-76
Taxation
  allowances, 263

Taxation (cont'd)
  business, 261-262
  correspondence, 260-261
  dates due, 261
  entertainment, 263
  generally, 259-268
  rates, 263-265
Temporary screens, 173
Tiling
  floor, 86-88
  roof, 62-65
  wall, 88-90
Tools and equipment hire, 207-216
Training and Enterprise Councils
  240
Traps, 99-100

Underpinning, 168-173

VAT, 242, 266-268
Vehicles, 262
Vinyl tiling, 87-88

Wallpapering, xxi, 146-147,
  186-187, 201-202
Wall tiling, 88-90
Waste pipes, 96-97
Water bars, 61
Welding and generators, 215
Weights of materials, 222-226
White patterned glass, 119-121
Windows
  hardwood, 51
  PVC-U, 51
  softwood, 50-51
Woodwool slabs, 48